A GEOGRAPHY OF
earth form

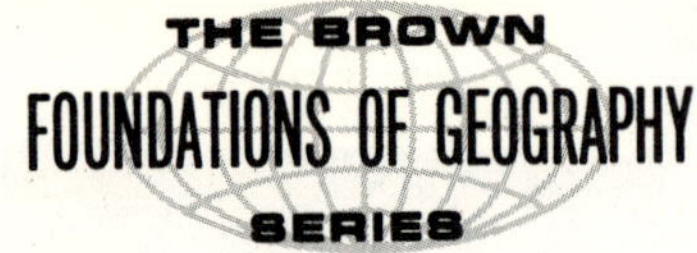

Consulting Editor
ROBERT H. FUSON
University of South Florida

A GEOGRAPHY OF

Agriculture
James R. Anderson, University of Florida

Soils
Robert M. Basile, University of Toledo

Transportation and Business Logistics
J. Edwin Becht, The University of Oklahoma

Plants and Animals
David J. de Laubenfels, Syracuse University

Geography
Robert H. Fuson, University of South Florida

The Atmosphere—(Second Edition)
John J. Hidore, Indiana University

Politics
W. A. Douglas Jackson, The University of Washington
Edward F. Bergman, Herbert H. Lehman College

Regions
James R. McDonald, Eastern Michigan University

Population and Settlement
Maurice E. McGaugh, Central Michigan University

Industrial Location
E. Willard Miller, The Pennsylvania State University

Water
Ralph E. Olson, The University of Oklahoma

Earth Form—(Second Edition)
Stuart C. Rothwell, University of South Florida

Landforms
John H. Vann, California State College, Hayward

Minerals
Walter H. Voskuil, University of Nevada

SECOND EDITION

A GEOGRAPHY OF
earth form

PREFACE TO PHYSICAL GEOGRAPHY

STUART C. ROTHWELL
University of South Florida

WM. C. BROWN COMPANY PUBLISHERS
DUBUQUE, IOWA

FOUNDATIONS OF GEOGRAPHY

Consulting Editor
Robert H. Fuson
University of South Florida

Geography is one of man's oldest sciences, yet it is as new as the Space Age. Knowledge of the earth obtained from satellite photography and measurement, remote sensing of the environment, and by means of other sophisticated techniques are really but a stage in the evolutionary process that began with ancient man's curiosity about his surroundings. Man has always been interested in the earth and the things on it. Today this interest may be channeled through the discipline of geography, which offers one means of organizing a vast amount of physical and cultural information.

The **Brown Foundations of Geography Series** has been created to facilitate the study of physical, cultural, and methodological geography at the college level. The **Series** is a carefully selected group of titles that covers the wide spectrum of basic geography. While the individual titles are self-contained, collectively they comprise a modern synthesis of major geographical principles. The underlying theme of each book is to foster an awareness of geography as an imaginative, evolving science.

Preface

This book is written as an introduction to an elementary college-level course in physical geography. Its purpose is to provide a minimal factual base upon which later studies of the geographical distributions of physical earth surface phenomena may be built. More specifically, it has a concern with the earth's size and shape and their relationships to matters of geographical interest. It deals with two general problems. Primarily, the earth's curvature is related to the two-fold problem of devising a locational system and representing earth surface data on maps. Secondarily, the earth's size and shape are related to other solar system facts in explanation of the geographical distribution of insolation and our time and locating systems. The book is largely a factual representation, with a minimum of geographical interpretation. No attempt is made to interpret the facts presented in their relationship to the natural processes which shape the details of the earth's surface.

The material is organized around the central idea that the earth may be thought of in either of two ways. Our planet is an object unique in itself. It is also a member of a system within which it reflects the behavior and attributes of other members of the system. The former view is developed in Part I, *The Earth: Basic Facts and Concepts,* which deals with the earth as an elemental particle in space; a thing apart from all else. As such the earth has form and dimensions, attributes which must be taken into account when a locational system is referred to or an extensive mapping project undertaken. Part I closes with chapters on maps and map projections.

Part II, *The Earth in the Solar System,* considers the earth in its relationships to its primary, the sun. In this case the earth is thought of as a dynamic element in a systematically organized grouping of interdependent and interacting bodies. As such it moves, and as it moves its motions determine the seasons, while providing the basis for our systems of time and location.

The author is indebted to several persons at the University of South Florida for their generous help given during the preparation of this manuscript. Dr. Robert Fuson of the Department of Geography and Professor Joseph Carr of the Department of Astronomy read the text and offered many constructive criticisms. Mr. Robert Oakes, undergraduate student in Geography at the University prepared the illustrations. To these colleagues, and to the many students whose questions have prompted the writing of this book, the author expresses his thanks. Any errors of fact or interpretation, however, are the author's own and he bears the responsibility for them.

Stuart C. Rothwell

Contents

Part I

The Earth:
Basic Facts and Concepts

The earth, in its form and dimensions, has long attracted the interest of man. The ancients, wondering about its shape, concluded that it had to be spherical and offered philosophical proofs. The earth's roundness was a necessary corollary to its assumed perfection. Others, at a later time, offered astronomical proofs. Still others, asking questions about its size, developed a practical mathematics and the instrumentation to measure it. For later man it became an object of study, not only for its own sake, but for more practical purposes as well. In either case, however, man's knowledge of the size and shape, or figure, of his earth has affected his concepts of distance, direction, location, and area. From these he developed his surveying, the science of earth measurement, and his cartography, the art and science of representing the rounded earth surface on a flat map.

The Earth's Figure

The earth's figure (its overall size and shape) may be described with varying degrees of generalization or precision to suit equally varying purposes. To illustrate, let us consider the actual topographic surface upon which the earth's measurements are taken. This surface varies widely in elevation and slope from place to place. Consequently, it does not lend itself to a mathematical description of the planet as a whole. Its numerous irregularities would require a prohibitive number of calculations. In practice then, when it is viewed as a whole, the earth is generalized as a smooth-surfaced body that can be mathematically described over wide areas. Such a smooth surface is assumed for each of the several three-dimensional forms commonly used to represent the earth's true figure. These forms (the *sphere*, a variety of *ellipsoids* or *spheroids*, and the *geoid*) vary in complexity and the ease with which they can be mathematically defined. Each is limited in the accuracy of its detail, and each has a distinct kind of usefulness.

The sphere is the simplest of the three basic forms. Although it is no more than a crude approximation of the true figure of the earth, it is quite satisfactory for a number of purposes. Precision is improved by representing the earth as a modified sphere, the more complex *oblate spheroid* (ellipsoid). This figure is obtained by flattening the spherical model at the poles and bulging it out at the equator to conform to known differences in the polar and equatorial diameters. The final precision is achieved by using the highly complex geoid, the "earthlike figure." This is the figure that would be formed by a surface which coincides with mean sea level over the oceans and mean sea level as it would be if it were extended under the large land masses. The geoidal surface is an irregular or undulating one which rises to known heights above, and dips to known levels below the surface of an ellipsoid. It provides a mathematically refined base to which points on the earth's surface may be referred. The geoid is, in itself, unique in the universe.

Notwithstanding its geodetic significance, the geoid has little relevance to a general geography, for it provides detail that is not required. Its principal use is in highly accurate surveying, for mapping purposes, or for artificial satellite and missile guidance work. The precise definition of the geoid is the task of *geodesy,* that branch of applied learning which occupies itself with the detailed measurement of the earth as a whole. Our prime concern here is with the sphere and its companion the ellipsoid, which, as representations of the earth, have much to do with a variety of factors of basic geographic interest. Our common systems of

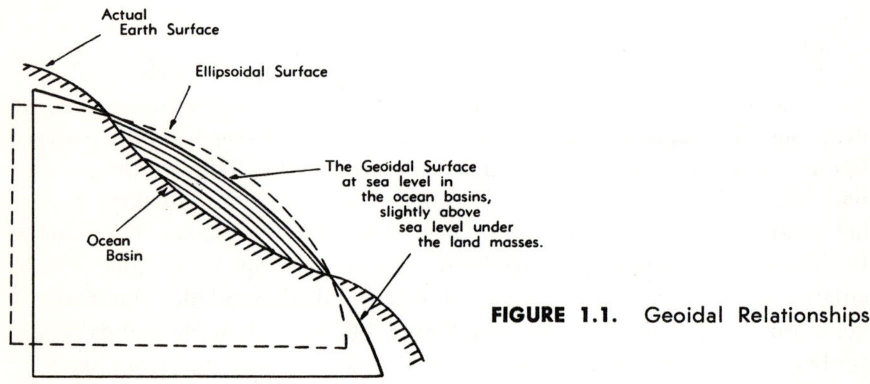

FIGURE 1.1. Geoidal Relationships

location and our measures of distance, area, and direction are all in some manner related to them. As mathematically defined models, the sphere and the ellipsoid reduce the extensive and varied surface of the earth to a comprehensible, definable whole.

The Spherical Earth

Considered as a true sphere, the earth is a smooth-surfaced body with a constant diameter of about 7927 statute miles and a circumference of 24,901 miles. Its total surface area is some 197,300,000 square miles. Strictly speaking, these data are only approximations; in reality both the polar diameter and the surface area are somewhat less than the figures cited. Furthermore, the surface is not smooth, but rather uneven. The deepest part of the oceans is some 7 miles below, and the highest point of land is some 5 1/2 miles above, sea level. Nevertheless, from a comprehensive point-of-view the error in the idealized sphere is relatively quite small. On an 18-inch globe the difference between the equatorial and polar diameters is so small as to be unobservable to the unaided eye.

A spherical earth is easy to work with because it can be uniquely defined by reference to one dimension only. Either its radius, diameter, or its circumference can be used. Primarily, it is a convenient model which can be used in the description and analysis of the earth as a member of the solar system. For geographical purposes such analysis is usually confined to considerations of the earth-sun-moon relationships. The sphere also provides an easily defined and imagined surface to which the geographical grid of lines of latitude and longitude is fitted. As a final point, when more precise statements about the earth's dimensions and form are required, the earth sphere provides the basic figure to which an ellipsoid can be compared.

development of the spherical earth concept

To the Greeks of 3000 years ago the earth was a flat disc, surrounded by the great river ocean. This was the world of Homer—circular in outline and surmounted by the great hemispheric vault of the heavens. By the sixth and fifth centuries B.C., other ideas concerning its form had evolved. Anaximander considered a cylindrical earth suspended in a spherical universe, and Anaximenes offered the concept of a rectangular one. About 500 B.C., Pythagoras and his followers taught that it was a sphere. They reasoned that since the gods had made the earth it had to be perfect. Perfection implied the earth's sphericity simply because the sphere was the only perfect geometric figure. The sphere was perfect because all points on its surface were equidistant from the center, and it had no beginning or end.

The Pythagorean sphere had considerable acceptance in the educated Greek world, but it remained for others, notably Aristotle (ca. 350 B.C.), to demonstrate it as a fact. Demonstration involved astronomical observations as proofs: proofs which are as valid today as when they were first offered. Chief among them was the fact that the earth throws a circular shadow across the moon during an eclipse, and Greek geometry indicated that only a spherical shadowing body could account for it. A second proof was the observation that as one traveled along a north-south line there was uniform change in the height of the sun and stars above the horizon. A third proof, unrelated to astronomical facts, was that an outbound ship appears to sink below the horizon as its distance from the shoreline increases. If these observations did not conclusively prove the earth's sphericity, they at least demonstrated that the earth had a generally curved surface and tended toward the spherical shape.

Acceptance of the idea of the sphere was followed by attempts to determine its dimensions. Plato and Archimedes (ca. 400 B.C. and 250 B.C., respectively) estimated the earth's circumference to be the equivalent of 30,000 and 40,000 miles. Eratosthenes (ca. 200 B.C.), by employing elementary geometry, actually measured the earth with astonishingly accurate results. Although in the light of contemporary knowledge it appears as though his accuracy was due in part to compensating errors in measurements, his method was a sound one. The principles he established are still in use for some of the more sophisticated earth measurements. Eratosthenes' method relied on a comparison of the ground distance and the latitude difference between the cities of Syene (Aswan) and Alexandria in Egypt. By sun observations he determined the latitude difference to be 1/50 of a circle. Since the ground distance was 5000 stadia, this meant that the circumference of the earth was 250,000 stadia, a figure which in our terms has been estimated to be from 25,000 to 30,000 miles.

Other measurements of the earth were made by the ancients. Of major significance was the work of Posidonius (ca. 100 B.C.). His principles were much the same as those of Eratosthenes except that he used a star as a reference. He calculated an earth circumference of 240,000 stadia (about 24,000 miles). Another measurement yielded a circumference of 18,000 miles. Unfortunately, this was the one which gained acceptance and it was not until the fifteenth century that it was discarded for a more realistic one. Columbus used the 18,000-mile figure. One wonders if he would have attempted his voyage over a much larger sea. Further, since his earth was believed to be much smaller than we know it to be, it is easy to understand why he was so certain he had reached Asia. In the ninth century A.D., the Arabs measured the length of a degree of latitude on the ground and computed a circumference. Much of their work has been lost to us because of uncertainties relating to the sizes of their measures of distance. It has been estimated that their work gave a circumference of just under 26,000 miles. Other significant attempts to measure the earth were not made until some centuries later.

The Oblate Earth

That the earth is only approximately a sphere appears to have been first considered seriously during the latter half of the 1600's. Astronomical and gravitational observations taken at different places and different times indicated that it was, in reality, slightly oblate, and not the spherical perfection the Greeks had assumed it to be. The telescope had revealed the planets Saturn and Jupiter to be slightly flattened at their poles, and the question as to whether or not this was also true of the earth naturally arose. Additionally, there was strong evidence to indicate that the earth's gravity varied with latitude. When considered in the context of the physical laws governing the force of gravity, this variation also suggested an oblate earth.

The occasion for gravity observations took place in 1671. In that year, Jean Richer, a French astronomer, noted that a very accurate pendulum clock lost time in relation to the stars when it was moved from Paris, 49° N, to French Guiana, 5° N. At this low latitude, the pendulum swung too slowly and it had to be shortened to speed it up. The slowdown resulted from a decrease in the pull of gravity upon the pendulum. The gravity decrease was later explained by reference to Newton's general physical law which states that any two particles of matter attract each other with a force directly proportional to their masses and inversely proportional to the square of the distance between them. In part this means that as the distance between the centers of two bodies is increased, the gravitational attraction between them is decreased. In Richer's case, it meant that while he was in the Guianas, his clock pendulum slowed down because it was farther from the center of the earth than when it was in Paris. That is, the earth's radius at the low latitude was greater than it was at the high. In consequence, the pull of gravity on the pendulum was lessened at the low latitude, the pendulum swung more slowly, and the clock ran slow. Like the astronomical observation noted above, this explanation suggested that the earth was, in truth, an oblate spheroid. Flattening at the poles and bulging at the equator was later explained as the result of the earth's rotation and accompanying centrifugal force which operated to stretch the equatorial diameter. Gravity observations are used today in determining the dimensions of the undulations on the geoid.

The implications of the astronomical and gravitational observations were too compelling to ignore. The question of the true shape of the earth could be resolved by taking careful measurements: if the earth were oblate, its curvature along a north-south line would vary. Curvature would be greatest at low latitudes, near the equator, and least at the higher ones, nearer the poles. However, early measurements taken by Cassini in France indicated that although curvature did vary, it varied in such a way as to indicate a *prolate* earth, elongated along the polar axis. It was not until the period 1735-1744, that highly accurate surveys proved conclusively that the earth was truly oblate, that is, slightly flattened at the poles and slightly bulging at the equator. These surveys were made in Lapland, 67° N, and near Quito, in present-day Ecuador, 4° N.

All of the early attempts to measure the earth suffered from poor distance measurements. Eratosthenes is said to have gauged the distance from Syene to Alexandria on the basis of how long it took a camel caravan to travel the route. The Arabs measured with wooden rods and, in 1525, the French physician Fernel counted the turns of a carriage wheel of known circumference to gauge a linear distance equivalent to one degree of latitude. Linear distance measure has always been less accurate than angular measurement. In view of this kind of limitation on the accuracy of results, development and use of the triangulation survey by Snellius about 1615 was a great step forward. In his system all long distances were to be calculated from one short base line measurement and a number of measured angles. Thus, by placing principal reliance on angular rather than linear measurements greater accuracy was assured. After Snellius, his principles were used for all long distance work where great precision was needed. Picard, in 1670, used them in measuring the distance between Paris and Amiens. In this survey which was modern in many respects, he measured his short base lines with wooden rods, used a theodolite, and calculated with logarithms.

The basic principles followed in a triangulation survey are illustrated in Figure 1.2. The length of the base line *AB* is carefully measured, as

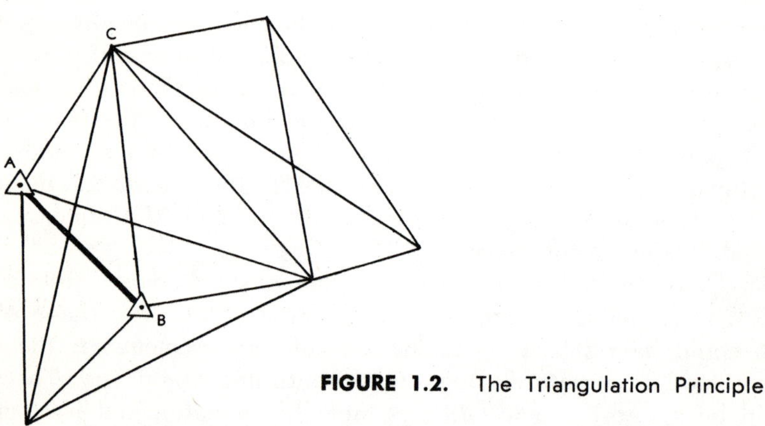

FIGURE 1.2. The Triangulation Principle

are the angles *CBA* and *CAB*. From these, the sides *BC* and *AC* may be calculated. The process of measuring angles and calculating distances may then be continued throughout the expanded net of triangles and quadrangles with no further need to measure distances except as a checking procedure. The results are highly accurate and, when used with precise modern instruments, the degree of accuracy is predictable. Convention has established that in normal surveying procedures there are three levels of accuracy. The *first order* delivers accuracy of one part in 25,000; the *second order,* one part in 10,000; and the *third order,* one part in 5000. In special work, accuracies of one part in 400,000 and one part in 1,000,000 have been achieved. Today, surveys of first order precision and higher are the basis of the mathematical definitions of the model ellipsoid.

For purposes of measurement and earth description, the flattened or oblate earth is expressed mathematically as an ellipsoid of rotation; a three-dimensional figure obtained by rotating an ellipse around its short diameter. An ellipse is a two-dimensional figure; a closed curve of such shape that for all points on it the sum of the distances from a point to two fixed points within the curve is constant. The two fixed points are the foci of the ellipse.

In contrast to the sphere which can be defined precisely by reference to one dimension only, the ellipsoid is described in terms of two dimensions, the major and minor axes which may be thought of as the long and short diameters. By convention, reference is made to the semi-axes (halves of the axes), and the flattening. Flattening is a measure of the departure from true sphericity. It is the ratio obtained by dividing the difference between the semi-major axis and the semi-minor axis by the semi-major axis, or:

$$\text{flattening} = \left(\frac{a\text{-}b}{a}\right)$$

Where: a = the semi-major axis
b = the semi-minor axis

By way of comparison, in a sphere *a* and *b* would be equal. The flattening would then amount to zero. On the opposite extreme, a flat plane would have no *b* dimension and its flattening would have a value of one. From this it follows that low value flattenings indicate near-sphericity,

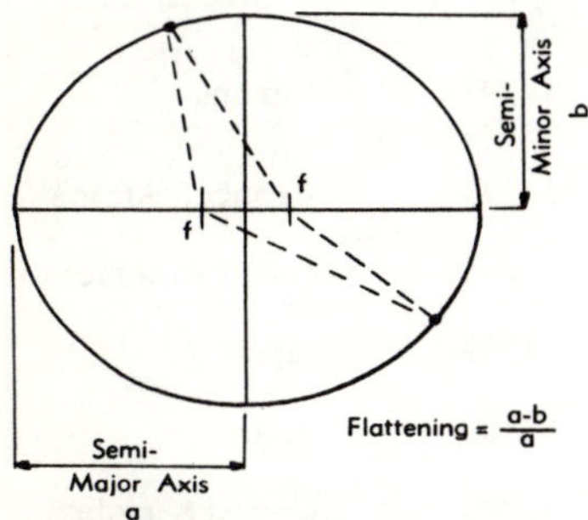

FIGURE 1.3. Elements of the Ellipse

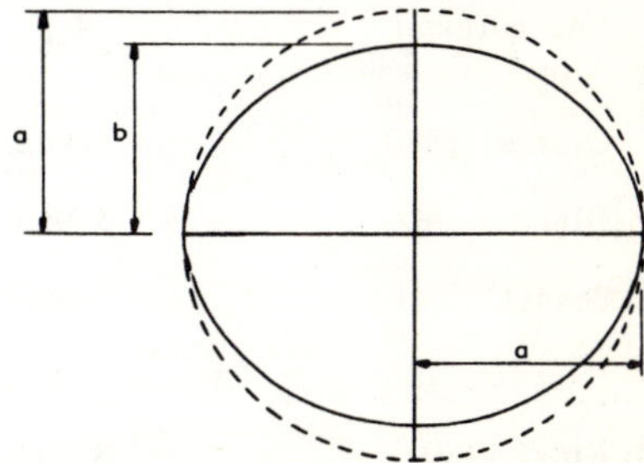

FIGURE 1.4. The Circle and the Ellipse Compared

while high values would indicate large departures from a precise spherical shape.

Earth ellipsoids are idealized mathematical models selected to describe approximations of the earth's form. They may be calculated for any part of the earth from measurements of the earth's curvature taken at the place in question. They differ from one another to the extent that the measured curvature differs from place to place. No single ellipsoid provides completely adequate representation for the whole of the earth simply because earth curvatures are so variable. For this reason, a figure known as the *ellipsoid of reference* has been calculated for each of several large areas of the world. Such an ellipsoid provides the surface upon which the geoid is produced and is the basis for precision mapping and long distance measurement in the area for which it has been developed.

Several ellipsoids of reference, as listed in Table 1.1, are currently in use in the areas indicated. Of these, Hayford's International, and Clarke's of 1866, are of particular interest. The first, computed by Hayford in 1909, was adopted by and recommended for general use by the International Union of Geodesy and Geophysics in 1924. Its diameters of approximately 7900 and 7927 statute miles are often quoted as the definitive earth dimensions. In contrast, the Clarke ellipsoid of 1866, which is used by the U.S. Geological Survey, (U.S.G.S.) is based upon diameters of approximately 7894 and 7926 statute miles for the polar and equatorial dimensions respectively.

TABLE 1.1

Dimensions of Selected Ellipsoids of Reference

Name of the Ellipsoid	Equatorial Radius Statute Miles	Flattening	Where Used
Krassowsky, 1940	3,960.89	1/298	U.S.S.R.
International (Hayford, 1909)	3,963.35	1/297	Europe
Clarke, 1880	3,963.48	1/293	France, Africa
Clarke, 1866	3,963.26	1/295	North America
Bessel, 1841	3,962.80	1/299	Japan
Everest, 1830	3,962.67	1/300	India
Airy, 1830	3,959.83	1/299	United Kingdom

Source: Burkhard, R. K., Capt., U.S.A.F., *Geodesy for The Layman*, U.S.A.F. Aeronautical Chart and Information Center, St. Louis, Missouri. 1962. p. 8.

Location on the Earth

Location on the earth is a key factor in geography. In the broadest sense it is a statement of relative position; a description of the "where" of an object with relation to another. Since it is stated relatively, it is measured in terms of a distance and a direction from some referent such as a point or a line, or another object. Location may be expressed either graphically, as on a globe or a map, or in verbal and/or numerical terms. It may be described with varying degrees of generalization to suit different requirements.

There are many systems by means of which a location may be described. In practice a position on the earth is said to be relative or absolute, depending on the referent that is used and the extent of the frame of reference to which the position is related. For example, the absolute location of a point is given in terms of latitude and longitude. In this case the point is positioned within a reference framework which extends over the earth and which is tied to natural features, the poles. In contrast, relative location (as the term is used here) is a statement of the position of an object, area, line, or point in its relation to some local feature. Such a statement, of course, has purely local meaning, since it ties the feature being located to a purely localized context. Thus, a point on a property boundary may be defined in terms of its distance and direction from some prominent landscape feature conveniently located nearby. A large tree, a bend in the river, a fence corner, or the like may suffice. Although the referent in this case is highly arbitrary and the statement of location tends to become imprecise as landscapes change, this procedure has been followed where the *metes and bounds* system of property description has been used. Similarly, an area may be located by reference to its generalized spacial relationship to another area or perhaps some major earth surface feature. Thus, the High Plains of North America lie east of the Rocky Mountains, Canada is north of the United States, and the like.

Although statements of relative location have obvious inadequacies, they are highly useful in particular situations. A metes and bounds boundary reference can be easily obscured where rivers shift their courses, fence lines are removed, and other landscape features modified. Nevertheless, in newly settled and unsurveyed lands the metes and bounds system has had great utility. In colonial times it was in general use, and it remains today in many legally recognized property descriptions.

Other statements of relative location are equally useful in other ways. For example, it would be impossible to explain the dryness of the High Plains mentioned above without recognition of the mountains immediately to the west. Air descending from the highlands has little tendency to yield its moisture. It has a drying effect instead. These are natural facts which have general relevance to the whole of the plains region and lend to it a particular climatic character. The location of the plains in relation to the mountains, then, is a factor to be reckoned with if this part of the earth's surface is to be understood. Similarly, the fact that Canada lies to the immediate north of the United States is a situation which has most certainly been significant to United States-Canadian relations through the years. Consequently, a statement of the relative positions of the two countries helps to explain such features as trade between them, their political alignments, and the like. Such a statement may also help in the understanding of any tensions that might exist between the two.

An earth surface feature may also be located in its relationship to a *grid.* A grid is a standardized system of points and lines. Within it the relative locations of all points are known and can be stated with precision. It is so designed as to have sufficient scope and detail to include all positions within the area it covers with the degree of precision required. Point locations are determined by the intersections of lines, or ordinates, which trend in different directions. Values or other identifying symbols are assigned to the ordinates, and a point location where two ordinates intersect is described in terms of point coordinates. An orderly arrangement of ordinates with a permanent and fixed point of origin is a grid system. The net-like pattern of its lines is referred to as the grid.

Many kinds of grid systems are currently in use. Each is suited to a specific purpose in a particular area; each has unique qualities in respect to its point of origin, the scheme by which its ordinates are identified, and the accuracy of its detail. The most basic, which may be thought of as providing a system of absolute location, is the geographic grid. Its ordinates are the curved lines of latitude and longitude, and its ordinate designations are stated in terms of angular measure. The origin is the intersection of a meridian, a north-south line, and the equator, an east-west line.

The geographic grid is first produced on the sphere, after which it is developed to suit the plane surface of a map. Since in most cases its angular measurements are difficult to deal with on the flat map, other grids which are keyed to some kind of linear measure are often super-imposed upon it. An added grid serves as an interpretive medium by means of which distances expressed as degrees of arc along lines on the earth's surface are translated into terms of yards, meters, and the like. When such a grid is used, point designations stated in degrees of latitude and longitude on the map form of the geographic grid may be read directly from the superimposed grid in linear terms.

The Geographic Grid

As noted above, the geographic grid provides the basis for all statements of absolute location on the earth's surface. It is fitted to the sphere, which it encompasses completely, and it is the all-inclusive system to which all other grids, points, lines and places may be related. Since its ordinates, the lines of latitude and longitude, encircle the earth, they are in fact circular. Their principal attributes as circles are discussed below.

In a mathematical sense there are two kinds of circles, great circles and small circles, which appear on the geographic grid. They are described as the lines formed at the intersections of planes and the earth sphere, for any plane passed through a portion of the sphere will intersect the sphere along a circle. If the plane passes through the center of the sphere, the circle formed is a great circle which is the largest one that can be drawn on the sphere. In this case the radii of the circle and the sphere are equal. In all other cases, where the plane does not pass through the center, the circles of intersection are small circles, with radii less than that of the sphere. Small circles range in size from the very smallest, almost a point, to the largest, with a diameter just under that of the great circle.

great circle characteristics

Aside from their functions as lines on the grid, great circles have considerable utility. Their unique properties make them highly useful in exposition regarding the earth's surface, as well as in the determination of location. Three such properties, along with some indications of their usefulness, are discussed in the paragraphs immediately following.

The shortest surface distance between any two points on the earth lies along an arc of a great circle. This distance is known as the great circle distance, and many air and ocean routes follow the great circle

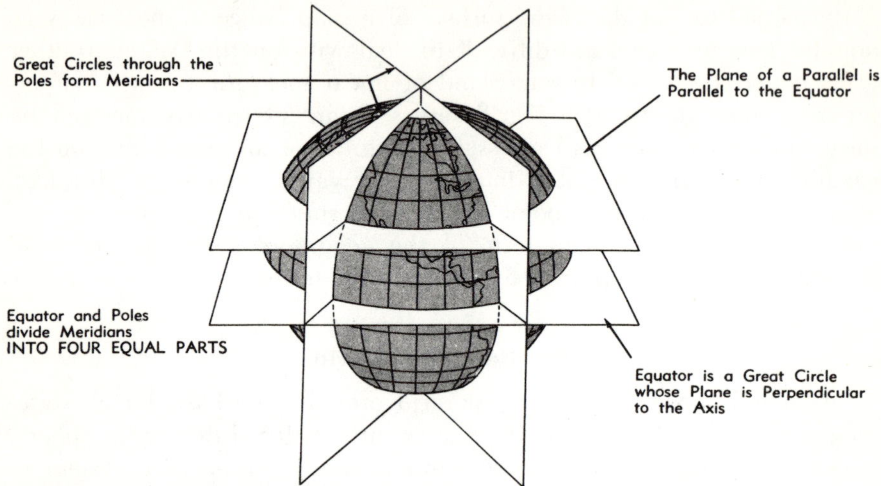

FIGURE 2.1. Planes and Circles of the Earth. (Courtesy of United States Air Force, from AFM 51-40, Vol. 1, Air Navigation.)

route because of its directness. In a sense the great circle on a sphere is analogous to a straight line on a plane surface. The straight line, too, is the shortest distance between two points. In another equally significant way, however, the two are very much unalike. The straight line on a plane is always a line of constant direction. In contrast, direction along a great circle is constant only when the great circle is the Equator or when it passes through the poles (see Figure 2.4).

Great circles bisect the earth and thus separate hemispheres. The circle of illumination which separates the sun-illuminated half of the earth from the dark half is such a circle, as is the Equator (see Figures 6.1 and 6.3, Chapter 6). In our system of latitude the Equator has a value of zero and separates the Northern from the Southern Hemisphere.

Any two great circles intersect and bisect one another. Referring back to the preceding paragraph, the circle of illumination intersects the Equator and, bisecting it, divides it into two equal arcs. This means that at any time of the year, one half of the Equator is experiencing day while the other half is experiencing night. This means further that the lengths of day and night on the Equator are always equal.

meridians, meridian circles, and parallels

Two special kinds of circles, meridian circles and circles of latitude, make up the geographic grid. They are differentiated from one another on the basis of their orientations with respect to the earth's axis and their functions in describing direction, and are so arranged as to intersect one another at right angles. The intersections mark points on the grid. Since the number of circles may be infinite, the possible number of circle intersections, or points, is also infinite. Any point on the globe, then, can be located at some intersection within the grid.

The earth rotates on its axis and in doing so provides two basic points of reference from which the circles of latitude, or parallels, may be positioned. These points, the geographic poles, are located at the ends of the axis of rotation where it intersects the sphere. Between them are the parallels of latitude, circles on the sphere which are formed by passing planes through it at right angles to the axis. The largest of the latitude circles, the Equator, lies midway between the poles and is a great circle. All of the others are small circles. As their name implies, all parallels are parallel to one another and the distance between any two of them is constant throughout their lengths. Parallels are true east-west lines.

In contrast to the parallels, the meridian circles are neither referenced to nature-given points nor are they parallel. They are all great circles formed by passing intersection planes through the length of the earth's axis. Consequently, all meridian circles intersect one another at the poles and are bisected at those points. The semicircles thus formed are known as meridians, true north-south lines which are positioned apart from one another at the Equator but which converge at the poles. Since meridians trend north-south, and parallels run east-west, they intersect at right angles on the globe.

latitude and longitude

Any point on the globe may be accurately located by reference to its geographic coordinates, or its latitude and longitude. Latitude is an arc distance measured in degrees north and south, away from the Equator. It is measured along a meridian and between the Equator and another parallel. Longitude is an arc distance measured in degrees east or west, away from some preselected first, or prime, meridian. It is measured along a parallel between two meridians.

Arc distances are equal in value to central angles of the earth sphere. Central angles are formed by radii that meet at the center of the sphere[1] and are measured in terms of degrees. In the case of latitude measurement the central angle lies in the plane of a meridian circle. Its sides, the

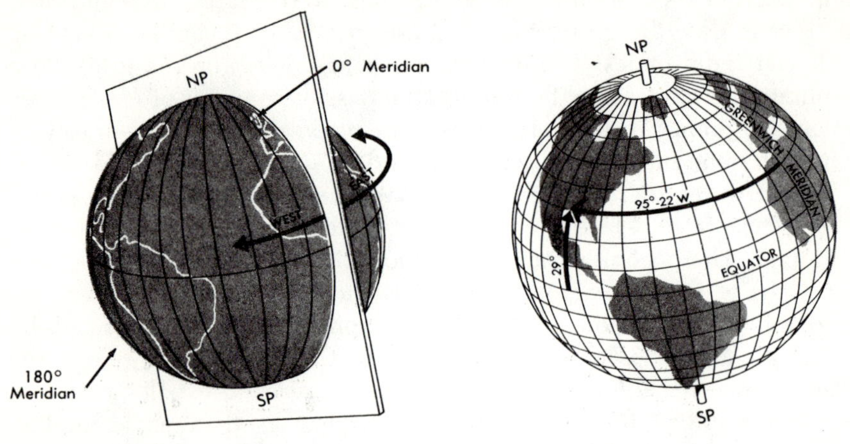

Longitude is Measured East and West of the Greenwich Meridian

Latitude is measured with Reference to the Equator; Longitude with Reference to the Prime Meridian

FIGURE 2.2.   Latitude and Longitude. (Courtesy of United States Air Force, from AFM 51-40, Vol. 1, Air Navigation.)

radii, intercept the arc distance on the circle. In the case of longitude the central angle lies in the plane of a parallel of latitude and intercepts an arc distance on the parallel circle.

In the generalized scheme of the geographic grid, zero values of latitude and longitude have been assigned to the Equator and a prime meridian respectively. For latitude, values range upward from 0° at the Equator to 90° north or south at the poles. With longitude, values increase from zero at the Prime Meridian to 180° east or west at the opposite half of the circle of the Prime Meridian.

Latitude values are derived from nature-given references. As previously explained, the poles are at the ends of the earth's axis and the equatorial great circle is formed by a plane which intersects the center of the

---

[1]This statement is valid only when the earth is assumed to be a true sphere. It is not valid when either an ellipsoid or the geoid is under consideration. In the latter case, the radii, or axes, meet at some point which is *not* at the center of the figure.

axis in a right angle. This means that there are 90° of latitude between the Equator and the poles, in complete accord with the range and the positions of the latitude values on the grid. In contrast, longitude values have been consistently positioned in quite arbitrary fashion. The prime meridian has often been changed, in large part as a matter of convenience. Since the times when location was first becoming systematized over the earth, many meridians have served as prime meridians in the longitude system. Quite commonly a prime meridian was defined as the meridian which passed through some key point in a capital city. One of the first to be used, the Meridian of Ferro, passed through the most western part of the Canary Islands. It was chosen so that all European longitudes were in the same hemisphere.

The prime meridian in greatest use today is the Greenwich Meridian. It was selected by international agreement in 1884 because it passed through the Royal Observatory located at Greenwich, near London, England. Some others now in use are listed below, and are located in relation to the Greenwich Meridian at the longitudes indicated.

TABLE 2.1

Locations of Prime Meridian Cities

Prime Meridian	Longitude, East of Greenwich
Rome, Italy	12° 27'
Athens, Greece	23° 42'
Istanbul, Turkey	28° 59'
Moscow, U.S.S.R.	37° 34'
Tokyo, Japan	139° 44'

The data indicate the corrections that would have to be made if longitude based on the Greenwich Meridian were changed to longitude based on some other. In the case of Istanbul, all west longitudes would be increased by 28° 59', and all east longitudes decreased by a similar amount. In no case, however, would any longitude be greater than 180°.

When using the geographic grid the absolute location of a point is completely defined in the following manner:

Latitude 27° 12' 42″ north, Longitude 102° 06' 09″ east of Greenwich

This means that the point in question is the intersection of a parallel of latitude located 27° 12' 42″ north of the Equator, and a meridian of longitude which is positioned at 102° 06' 09″ east of the Greenwich Meridian.

Antipodal points (from the Greek, *anti,* meaning against, and *podes,* meaning feet) are points on the globe which lie opposite to one another. Thus, the *antipode* of the point in question above is located at Latitude 27° 12′ 42″ south, and Longitude 77° 53′ 51″ west.

comparisons of arc lengths on the geographic grid

If the earth were a perfect sphere the meridian circles would be true mathematical circles and the length in miles of a degree of latitude would be everywhere the same. However, the earth is not a true sphere, but an ellipsoid which is slightly flattened at the poles. Thus, the meridian circles have less curvature near the poles than near the Equator. In consequence, the linear distance corresponding to one degree of latitude along a meridian increases from the Equator to the poles. The linear length of a degree of longitude, too, changes between the Equator and the poles. It decreases from a maximum at the former to zero at the latter. This, of course, results from the fact that longitude is measured between meridians and meridians converge at the poles.

The linear lengths of degrees of latitude and longitude as they are measured or computed at selected latitudes are given in Table 2.2. Comparative data for the Clarke ellipsoid of 1866 (used for mapping by the U.S.G.S.) and a sphere of constant radius of approximately 3963 statute miles are listed. Note that this radius is equal to the equatorial radius of the Clarke ellipsoid as it is given in Table 1.1.

The table gives the linear lengths of 1° arcs of latitude and longitude at selected latitudes on the sphere and the ellipsoid, and shows the dif-

TABLE 2.2

COMPARATIVE ARC-LENGTH DATA

The Clarke Ellipsoid of 1866, and a Sphere of Radius of 3963 Statute Miles

Location: Latitude	One Degree of Longitude Statute Miles			One Degree of Latitude Statute Miles		
	Ellipsoid	Sphere	Difference	Ellipsoid	Sphere	Difference
0	69.17	69.17	0.00	68.70	69.17	0.47
10	68.12	68.11	0.01	68.72	69.17	0.45
20	65.02	64.99	0.03	68.78	69.17	0.39
30	59.95	59.90	0.05	68.87	69.17	0.30
40	53.06	52.98	0.08	68.99	69.17	0.18
50	44.55	44.40	0.15	69.11	69.17	0.06
60	34.67	34.58	0.09	69.23	69.17	0.06
70	23.72	23.65	0.07	69.32	69.17	0.15
80	12.05	11.97	0.08	69.38	69.17	0.21
90	0.00	0.00	0.00	69.40	69.17	0.23

Source: Data for the ellipsoid from: Johnson, W. E., *Mathematical Geography,* New York, 1907, p. 44.

ferences between the lengths of corresponding arcs on the two figures. When compared, the lengths of the corresponding arcs are seen to vary unequally from the Equator to the poles. The small differences in corresponding arcs indicate the validity of using the sphere, rather than the more realistic ellipsoid, as the basis of the geographic grid.

A first comparison shows that the number of statute miles per degree of latitude is a constant on the sphere. In contrast, the number of miles per latitude-degree on the ellipsoid changes from 68.70 at the Equator to 69.40 at the poles. The rate of change at intervening latitudes, however, varies. This is demonstrated when the linear measures of arcs located at the same latitudes on the sphere and the ellipsoid are compared. Differences in the lengths of 1° arcs change by decreasing between the Equator and latitudes 50° and 60°, and by increasing from there toward the pole. This means, of course, that the ellipsoid and the sphere are very nearly coincident at the higher middle latitudes. The degree of coincidence is least at the Equator. Here the difference between the lengths of corresponding 1° arcs is 0.47 of a mile, somewhat less than 1 per cent of 69 miles.

In the case of the 1° arcs of longitude their linear lengths decrease toward the poles on both the ellipsoid and the sphere. They are seen to be different, however, when arc segments of the ellipsoid and the sphere are compared. At latitude 50° the disparity between them is 0.15 of a mile, or about 0.3 per cent of 45 miles. At latitude 80° their difference is 0.08 of a mile, or about 0.6 per cent of 12 miles.

The comparisons demonstrate that differences between the ellipsoid and the sphere are relatively quite small. In no case does the disparity between corresponding arc lengths exceed 1 per cent of the lengths of the arcs. For this reason, as well as the additional one that computations on the ellipsoid are quite lengthy, general cartographic work commonly disregards the differences. It uses the dimensions of the sphere in its flattened representations of the geographic grid, considering them to be adequate approximations of the more realistic ellipsoid. For more demanding work, such as the calculation of the geoid and highly precise mapping, ellipsoid dimensions are a requirement.

latitude and longitude relationships

On the sphere, as shown in Table 2.2, the lengths of 1° arcs of latitude and longitude are equal at the Equator. At other latitudes, the longitude arcs decrease systematically as follows:

Length of 1° Longitude = Length of 1° Latitude (cosine X°) where:
X° is the latitude of the longitude arc being measured.

This principle is commonly used to derive the longitude dimensions which are employed when the spherical earth grid is flattened so that it may be used in constructing a flat map. It is the fundamental formula for general cartographic work.

Understanding of the mathematical relationships between arc distances and other elements of the sphere is as old as the geometry of the ancients. It was necessary to Eratosthenes' measurement, as well as to all subsequent efforts to determine the earth's form and dimensions; it is implicit in the most widely used systems of linear distance measurement today.

linear measurements on the geographic grid

In the United States linear distances are now measured in terms of either the Metric or the U.S. Customary (English) system, with the meter considered to be absolute standard. The meter also appears as the basic unit of measurement on the U.T. Mercator Grid, a coordinate system which is superimposed upon many of the maps of the U.S.G.S. Its principal use, however, is in scientific work where either meters or kilometers are commonly used in expressing earth dimensions.

The meter was originally defined in terms of earth dimensions. In 1791, on the recommendation of the French Academy of Sciences, it was defined in France as 1/10,000,000 of the supposed meridional arc distance between the Equator and the Pole along the Paris meridian. Later, when it became possible to measure more precisely, the total distance was found to be 10,002,286 of these meters, and a re-definition was necessary. The standard meter then became a distance marked off on a platinum-iridium bar held at a temperature of 0° Centigrade. In 1960, to insure a permanent accuracy of a high order, it was again defined in terms of the wavelength of orange-red light emitted by electrically charged krypton gas.

In the U.S. Customary system the yard serves as the counterpart of the meter. In 1832 it was defined by law as a distance marked off on a brass bar held at a temperature of 62° Fahrenheit. Effective in 1959, a standard yard of exactly 0.9144 of a meter was agreed upon by the governments of the United States, Australia, Canada, New Zealand, South Africa and the United Kingdom. From this comes the U.S. Survey foot of 1200/3937 meters, a measure applied in all geodetic surveys where the foot is used.

Other measures which are commonly used to describe long distances include the various kinds of miles. In the United States land distances are measured in *Statute Miles* which are defined by law as equal to 1760 yards, or 5280 feet. Among navigators the *Nautical Mile* has been a standard for centuries and, like the meter and the yard, has undergone several re-definitions. In earliest times it was considered to be the meridional arc distance equal to 1 minute of latitude. Since this was known to vary, the several maritime nations adopted different standards. Subsequently, in a search for uniformity, it was defined as the length of 1 minute of arc on a great circle of a sphere which had a surface area equal to that of the earth. As used in the U.S. Navy it was equivalent to 6080.27 U.S. feet. In 1954, the Departments of Defense and Commerce adopted the International Nautical Mile of 6076.10 feet. A brief statement of equivalents is given below:

One old nautical mile equals	One International Nautical Mile equals
6080.27 U.S. feet	6076.1 U.S. feet
1853.25 meters	1852.0 International meters
1.1516 statute miles	1.15 statute miles

Another kind of mile, sometimes erroneously referred to as a nautical mile, is the *Geographical Mile*. It is equivalent to about 6087.08 U.S. feet, and is defined as the linear length of 1 minute of arc of the earth's Equator.

direction on the geographic grid

Direction is a statement of the position of a point on the earth relative to another, without any reference being made to the distance between them. It is not in itself an angle, but it is usually measured in terms of the angle between a reference line and the shortest line that can be drawn to the point of interest from the point of observation. When measured on the earth or on the grid, the reference line may be a true north-south line (a meridian), a line from the point of observation to magnetic north, or any other line of convenience. Since the line to the point of interest is the shortest possible, it lies on a great circle. More specifically, then, direction on the earth is indicated by the angle between the reference line and a great circle which intersects it at the point of observation.

Direction may be stated as an *azimuth* or as a *bearing*. As an azimuth it is the horizontal angle measured clockwise from the reference line. Its value may range from 0° to 360°. In expressing direction as a bearing, quadrants are referred to. Quadrants are 90° sectors, or quarters, of the horizon circle which is centered on the point of observation. They are defined by referring to their position within the horizon circle, that is, as the northeast, southeast, southwest and northwest quadrants. A bear-

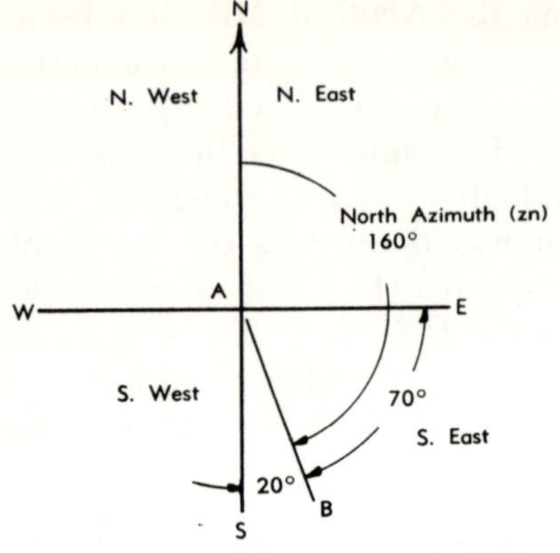

Bearing of Line AB, S20°E.
Azimuth of Line AB, 160°

FIGURE 2.3.   Bearings and Azimuths

ing is given as a horizontal angle within a quadrant. It is referred to a meridian, and is measured either clockwise or counterclockwise from either north or south. Its value may range from 0° to a maximum of 90°. As an example, a line with a true north azimuth of 160° lies 160° in a clockwise direction from the reference meridian, a line between the observation point and true north. It lies in the southeast quadrant of the horizon circle, and has a bearing of 20° east (or counterclockwise) from a south line. Its bearing is described as "South, 20° East." A line with an azimuth of 290° has a bearing of "North, 70° West."

When described on the globe any line used to bound a directional angle is taken as the shortest line between two points. It is, then, a segment of a great circle along which direction in relation to true north is constantly changing. This remains true in all cases except where the line is taken along a meridian or the Equator. Thus, true north azimuths and bearings are referred to lines which follow a north-south direction throughout their lengths (meridians). The line of observation, however, does not continue in an unchanging direction. This becomes clear when it is recalled that meridians converge at the poles. The line of observation intersects successive meridians at angles which become more and more different from the original angle as distance from the point of observation increases. As shown in Figure 2.4, a great circle segment drawn from New York to London will intersect the New York meridian at an angle of about 50°. At the point of intersection, it has an azimuth of 50°. Its bearing is North, 50° East. Near London, however, its azimuth is about 90°. Its bearing is due east. This means that a navigator plotting a great circle course from New York to London must continually plot new directions or headings if he is to continue on the great circle route.

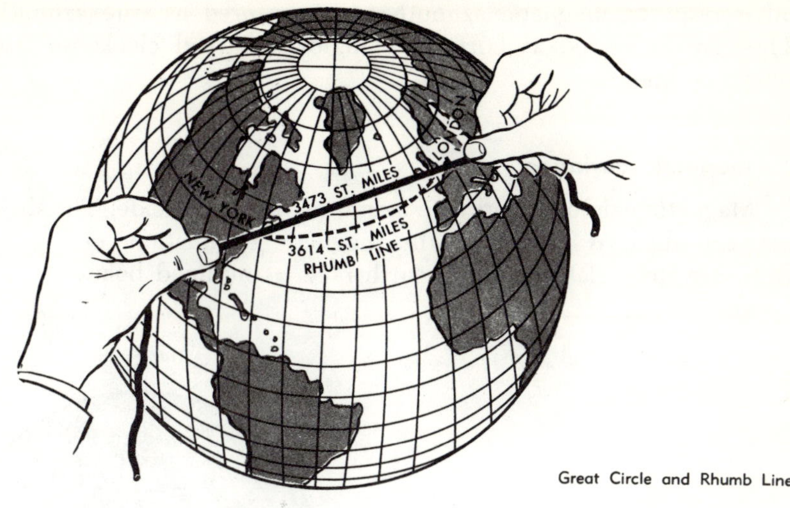

Great Circle and Rhumb Line

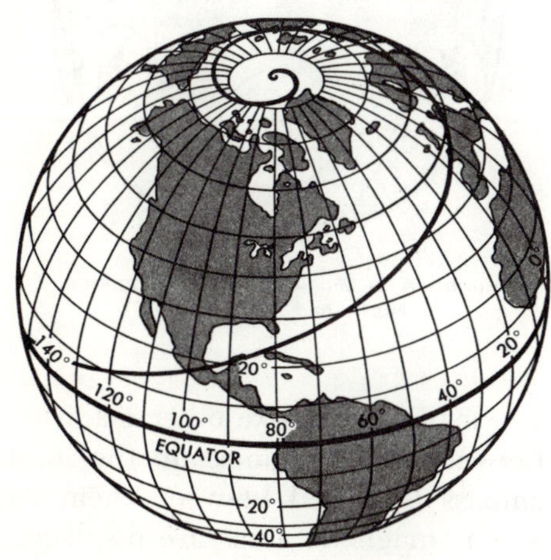

A Rhumb Line or Loxodrome

FIGURE 2.4. Great Circle and Rhumb Line Relationships. (Courtesy of United States Air Force, from AFM 51-40, Vol. 1, Air Navigation.)

A line that is drawn between points in such a manner that its direction is constant in respect to true north is known as a *rhumb line*. An aircraft following such a course would continue along a spiral curve toward a pole, and would, in theory at least, never reach it. The spiral described is called a *loxodrome*, or loxodromic curve.

Direction may also be expressed in relation to magnetic north. In this case it is referred to as a magnetic azimuth. The reference line is a small segment of a great circle along which the compass needle points. In other respects, magnetic azimuths are measured as true azimuths are. They are expressed as horizontal angles and read clockwise from the reference line.

magnetic variation

Magnetic azimuths and true azimuths rarely coincide. The difference between them is known as the magnetic declination, or variation, a feature which relates to basic earth facts as outlined below.

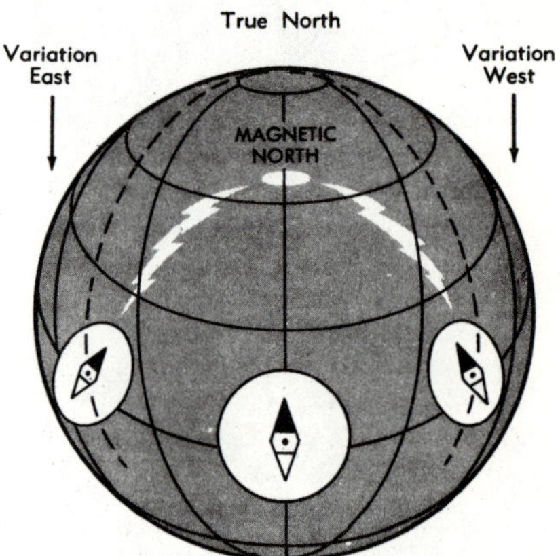

FIGURE 2.5. Magnetic Variation. (Courtesy of United States Air Force, from AFM 51-40, Vol. 1, Air Navigation.)

The earth may be likened to a huge magnet and, like other magnets, it has two magnetic polar regions. Between these regions are the lines of magnetic force. If a magnetic compass is placed between them its needle will align itself with the lines. The magnetic areas are displaced from the geographic poles around which the earth rotates. In consequence, the direction indicated by the compass needle at nearly all points on the earth does not coincide with a true north-south line. At any of these points the angle between the meridian and the line indicated by the compass needle is known as the magnetic declination, or variation. If the line of the compass needle lies to the west of the meridian, variation is said to be west. If it lies to the east, it is said to be east.

The magnetic variation varies from place to place, and from time to time at the same place. Its value ranges from 180° to zero degrees. Furthermore, the pattern of its variations over the face of the earth at any one time is extremely irregular. The magnetic declination is the

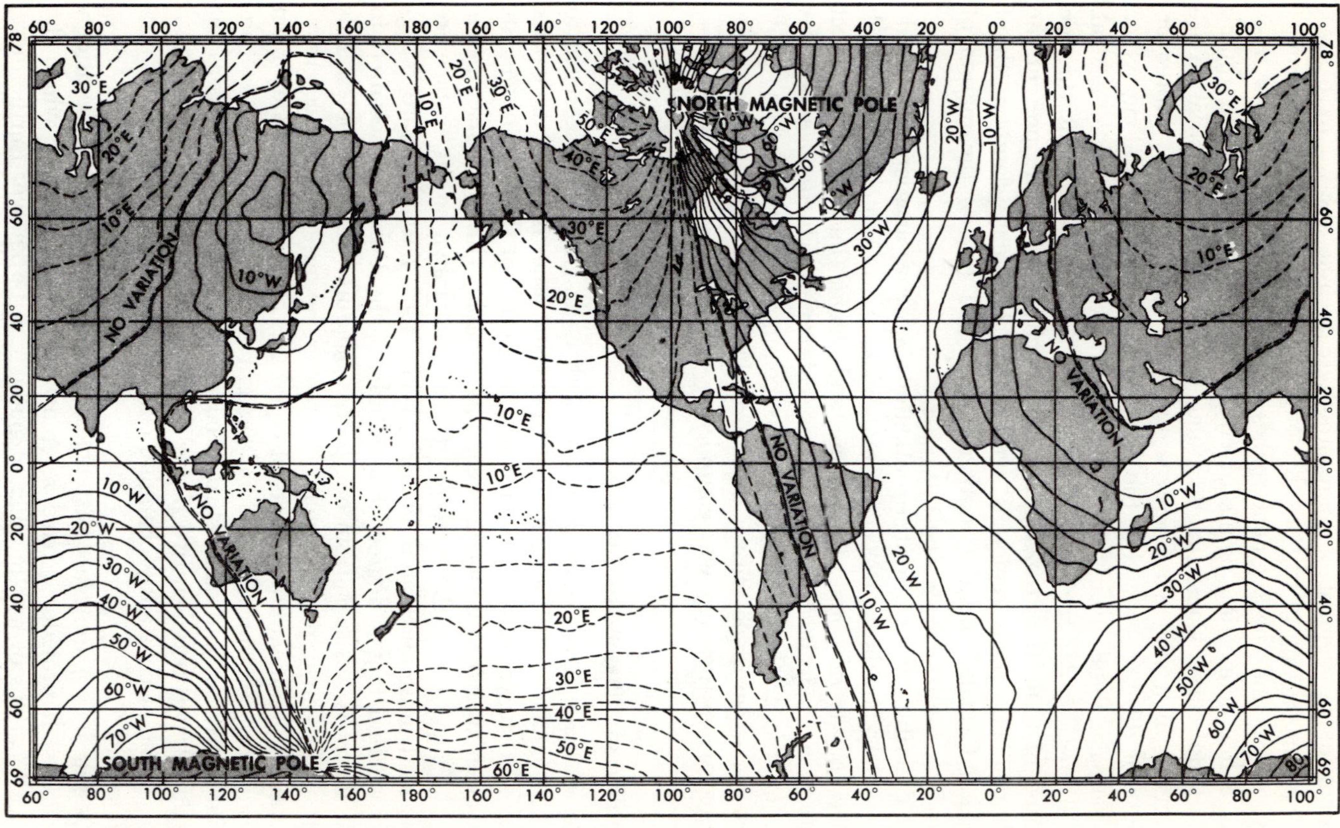

FIGURE 2.6. Isogonic Lines (Courtesy of United States Air Force, from AFM 51-40, Vol. 1, Air Navigation.)

direction of magnetic north from true north. It is measured between a meridian and a segment of some other great circle. As we have seen, directions in relation to true north along such a segment are different at every intersection the segment makes with a meridian. On this basis alone, a line connecting points of equal declination must be a rhumb line, a spiral curve. Consequently, points of equal magnetic variation are distributed in a somewhat uneven pattern over the earth. Moreover, because of differences within the earth, the lines of force along which the compass aligns itself are neither fixed, nor are they symmetrical. The magnetic polar regions themselves shift in their locations through a period of time, thus causing declination changes. At the present the north magnetic polar region is centered on Bathurst Island, north of mainland Canada, and the southern polar area lies on the margin of Antarctica, south of New Zealand.

In describing the world distribution of magnetic variation, small scale maps which show large areas commonly make use of *isogonic* lines which connect points of equal declination, and *agonic* lines which connect points of zero declination. On larger scale maps which cover smaller areas on the earth, the declination is commonly shown by an arrow which makes an angle with a true north line. Declination is shown for the year when the map was compiled.

utility of the geographic grid

As we have seen, location on the earth is described by reference to the distance and direction of a point from some other point on the earth's surface. What we have also seen is that measures of these two elements quite often appear to be inconsistent and variable. This, of course, results from the curvature of the earth's surface. Lines drawn on the surface are curves and must be treated as such. When their curvatures are recognized we find that a degree of longitude at the Equator, measured in linear terms, is different from one that is measured near the poles. This is a result of converging meridians. Moreover, because the surface curvature varies, the linear equivalents of equal arcs of latitude also vary from the Equator, poleward. Directional measures, too, are affected by converging meridians and earth surface curvatures. Thus, a line of constant direction drawn between two points on the globe is not straight. It is a spiral; a line of constantly changing curvature when viewed on the geographic grid.

Despite its apparent inconsistencies, the geographic grid has considerable utility. It provides a fixed referencing system by use of which the relative locations of any number of points can be known. Furthermore, by means which are described in a later chapter, the geographic grid can be flattened, albeit with some distortions, and used as the framework of the map. Proper use of the map, however, requires understanding of the effects of the characteristics of the geographic grid upon our designations of location on the earth. Methods of determining point locations on the spherical earth and the general principles which guide the techniques of their transfer to the flat map are described below.

Determination of Point Locations on The Geographic Grid

Point locations on the geographic grid are determined by methods of ground survey, either *geodetic* or *plane*. Geodetic methods are used to cover large areas. They produce highly realistic results, taking into account the size, curvature, and figure of the earth. Directions and distances, as measured on the ground, are referred to the ellipsoid and corrected to conform to it. Thus, a geodetic survey can be carried through long distances to tie widely separated areas together with a minimum of error. In contrast, plane surveying assumes the earth's surface to be flat. Direct measurements on the ground are treated as if they were taken on a plane. In consequence, plane surveys yield somewhat less accurate results than geodetic surveys. Their measured angles and distances depart more and more from reality as they are carried farther and farther from their starting points. As a general rule 100 square miles is considered to be about the largest area within which plane survey methods are acceptable. Beyond this errors become too great. When the permissible size of the surveyed area has been reached, a new starting point is selected and a new survey begun. In modern procedures the starting point of a plane survey is located by geodetic methods. At other times, prior to the development of elaborate geodetic networks, and today in areas where geodetic networks are not yet established, the plane survey is run from some prominent landmark, as in the metes and bounds system.

The ground survey produces the basic data for the mapmaker. It determines and designates point locations on the ground. The cartographer then uses the data to establish the relative positions of representative points on the map. Distances between point locations on the map are proportioned to measurements taken between the corresponding points on the ground. Where feasible, and where adequate data are available, points are located on a *graticule*, a scaled network of meridians and parallels which is drawn on a flat sheet of paper. As starting points for precise work the cartographer uses *geodetic control points* in his design of the graticule. These points have been previously located on the ellipsoid by the geodetic survey. The lines, points, and angles used to fill in between the control points have been defined and located by plane survey methods.

horizontal control

Horizontal control points are points where the latitude and longitude are known. They are established on the ellipsoid by triangulation or traverse survey techniques, with first- and second-order degrees of accuracy. As shown in Chapter 1, triangulation involves the development of a network of triangles and quadrilaterals, with a minimum of distance measurement. A *traverse* is quite different and is somewhat akin to dead-reckoning navigation. It involves the laying out of lines and measured angles between points along a general line of changing directions known as the traverse. The lengths of lines may be measured directly or triangulated. The traverse may be closed, in which case it is brought back to the starting point as a check, or it may be left open. In general, traverse work in a geodetic survey is of second-order accuracy and is used to fill in between primary elements in the triangulation network. Primary elements are located with first-order accuracy.

The field survey involves a preliminary positioning of two widely spaced points by astronomical observations. Base lines are laid out from the points, and a triangulation net developed between them. This procedure establishes an *arc of triangulation* within which the vertices of triangles are the first-order control points. A series of arcs which is so organized as to cover a definite area is known as a triangulation system. Areas between the arcs in a system are filled in by plane survey methods.

Horizontal control points are referred to as *triangulation stations*. They are permanently marked on the ground by bronze disks imbedded in concrete blocks, or monuments. As a general policy, the U.S. Coast and Geodetic Survey (U.S.C. & G.S.) establishes one monumented horizontal control point for every 50 square miles. This spacing permits the use of triangulation stations as starting points for local plane surveys. As noted previously, such surveys can be used within an area of 100 square miles or less.

vertical control

Vertical control points are points where the elevation of a marker above or below a standard datum is known. In the United States, the datum is the standard sea level of 1929. The points are established by vertical surveying, or leveling, in an operation which is usually carried out separately from the horizontal surveying activities. Like other control points, they are monumented. The bronze disk of the monument (*bench mark*) is stamped with the elevation on the top of the mark. In countries where the English system of measurement is used, and where the bench mark has been recently located, elevations are given in survey feet.

extent of the geodetic control network

In the United States the first geodetic surveys were carried out along the Atlantic Coast. They were subsequently extended westward in several long transcontinental arcs which were linked together by a series of shorter arcs running north-south. In 1913 the entire system was tied into one central control point located at Meade's Ranch, Kansas. In 1927 all stations in the system, as well as others in the Mexican and Canadian systems, were checked and related to the point. Meade's Ranch is now referred to as the North American Datum of 1927.

Other extensive systems are those of India, Australia, and the Philippines, and those which cover much of the U.S.S.R. and countries in Western Europe. A 7000 mile long triangulation arc has been run from South Africa to Scandinavia. Since World War II, several United States government agencies have cooperated with a number of interested foreign countries to carry out extensive surveys that will ultimately link all of the great systems together. To carry out this program, highly sophisticated techniques have been developed. Aerial surveys using electronic measuring devices and artificial satellite observations have been used to establish new triangulation stations, remeasure the ocean basins, correct errors in the locations of mountain peaks, and even relocate islands. Despite these achievements much of the world has yet to be adequately surveyed and monumented for geodetic control purposes. In general, control points are lacking in northern Canada, South and Central America, most of Africa, and the interior of Asia.

Map Projections

As we have seen from Chapter 2, point locations on the earth are specified by latitude and longitude designations taken from the spherical grid. They are recorded, however, on a flat graticule. It is impractical to record a survey and develop a map on any surface other than a plane. Thus, practical considerations require that the spherical grid be reproduced in some way as a flattened arrangement of parallels and meridians. This requirement is met by the systematic transfer of point locations from the sphere to a plane, with the transfer being made by mathematical and/or graphical means. From this, a *map projection,* or map graticule is produced. The map is developed by locating the lines and points which identify earth surface features within the projection. Thus, a map projection is properly defined as a systematic arrangement of lines of latitude and longitude on a plane, while the map itself is a representation of earth features developed around the projection.

Because the earth is a sphere, all map projections and maps have some distortions in them. A spherical surface cannot be flattened without introducing some stretching or compression into its dimensions. Consequently, the flattened arrangement of meridians and parallels that we call a projection will always have some error in the orientations and the dimensions of its lines. Error, however, can be controlled, and some kinds of it can be eliminated entirely by use of a particular kind of projection. The error, whatever its nature, will of course have a final effect on the map within which some kind of dimensional and/or directional inexactness must be expected.

Inexactness does not destroy the utility of the map. But it must be understood if the map is to be used effectively and honestly. Since the qualities of exactness or inexactness of the map are determined in large part by the fidelity of the projection in showing some aspects of the spherical grid truly, skill in the art of using maps requires some knowledge of the projections themselves.

Some Initial Considerations

All true projections can be defined mathematically. Mathematical operations alone can produce some; others can be produced by purely graphical techniques, while still others are developed by use of both graphic and mathematical methods. In the case of the first, formulas are used to transfer latitude and longitude values on the sphere to a grid system of plane coordinates. Points corresponding to these specified values are then positioned in their calculated relationships to one another on a flat map sheet. After all calculated points have been located on the sheet, the graticule is produced by drawing continuous curves through all of the points plotted for each meridian and each parallel.

In some cases similar results can be achieved by purely graphical techniques which produce the so-called *perspective,* or *geometric* projections. In these, points are projected directly to the projection surface, with no calculations. Lines are then drawn as noted above and the graticule completed. When both the graphical and the mathematical methods are used, a preliminary graticule produced by perspective means is modified by mathematical operations. Most maps of large areas are developed on projections of the latter type, although many can also be produced entirely graphically.

the perspective principle

If a wire model of the spherical geographic grid were constructed, a light shown on it, and its shadow thrown on a screen, the individual shadows of the lines of latitude and longitude would together form a pattern. This pattern would be a perspective map projection. If the light were moved, either toward or away from the center of the model, the pattern would change. The lengths and curvatures of lines would be altered and the relative positions of the line intersections, or points, modifed accordingly. These considerations establish the principle of the perspective projection. They demonstrate that many such projections of the geographic grid are possible.

In illustration, four projections which may be thought of as basic are shown in Figures 3.1, 3.2, 3.4 and 3.5 (pp. 37-40). Each figure shows two diagrams. To the left is the *generating globe,* the wire model through which the light rays pass. The line *AB* represents an edge of the *plane of projection.* It is drawn tangent to the globe at the intersection of 110° W and the Equator. The intersection is the point of tangency of the projection plane and the globe. The diagram to the right is the projection and its map viewed from directly above the point of tangency. Light rays are shown as originating at point *P*, a point of light located at different distances from each of the generating globes.

Other projection surfaces are the cone and the cylinder. When these are used, the light source is usually held to be at the center of the generating globe. They are illustrated in succeeding figures.

Quite clearly the location of the light source controls the positioning of lines and points on the projection surface, while the positioning of the lines and points on the projection controls the qualities of the representations of land and water areas. Each graticule is different from the others and the differences are reflected in the shapes and sizes of features on the maps.

Differences in the patterns of lines and points on the graticules suggest that flat projections do not show all points taken from the globe in their true locational relationships with one another. Mathematical considerations of the problem of transferring a three-sided figure from the sphere to a plane provide a proof.

the limitations of projections

As a preliminary to discussion of the mathematical limitations of the projection, it will be recalled that point locations are stated in terms of distance and direction. Consequently, if the true relative locations of points are to be transferred from the sphere to a plane, the lengths of lines and the angles between lines must be shown truly. The comparison of spherical and plane triangles given below points out that these conditions cannot be met on a single projection.

A closed, three-sided figure on a sphere is referred to as a spherical triangle. Its sides are great circle arcs which intersect each other at the vertices to form three spherical angles. Spherical trigonometry tells us that the sum of the angles in a spherical triangle is always greater than 180°, with the excess above 180° directly proportional to the area of the figure. This contrasts to a plane triangle drawn on a flat surface. In the plane triangle the sum of the angles is always equal to exactly 180°, regardless of its size. As a consequence of this difference no spherical triangle can be exactly represented on a plane surface; if the sum of the angles is held, the triangle must become a quadrilateral when it is placed on a plane, and, if the true lengths of lines are held, the angles must change. Now since point locations are determined by reference to both the lengths of lines and the angles between them, it becomes clear that when any three points are under consideration it is impossible to project them in their true relative positions from the sphere to the plane. Moreover, because the sum of the angles in the spherical triangle increases as the area of the triangle increases, the farther apart the points are on the globe, the greater the error in their positioning on the plane becomes.

Distortions, then, are built into all graticules and all maps that are developed on them. A spherical surface cannot be flattened without changing the distance-direction relationships between points on the surface. Consequently, on any projection some points must be misplaced. Since point locations determine the lengths and azimuths of lines, and ultimately fix the sizes and shapes of areas bounded by lines, any point misplacement on the projection is reflected in the map.

Map Projections and the Purposes of Maps

Distortions in map projections can be controlled. Some can be eliminated entirely. More importantly, a map projection can be designed positively. That is, a projection can be made so that its finished map will represent some aspect of the earth's surface truly. The aspect desired will, of course, be determined by the purpose of the final map.

Depending on circumstances, the purpose of the map may require that areas be represented properly in their true shape or in their true relative sizes. In other situations it may be required to show the true relative lengths or directions of lines. In either case some map projection has been devised to fulfill the requirements of the map and these requirements need to be looked into before a projection is chosen and a mapping project is begun. Choice of this kind is a cartographic skill that should be understood by mapmaker and map-user alike.

The choice of a projection that will fulfill the requirements of the map developed upon it is facilitated by a systematic review of the basic features of the geographic grid, as it appears on the sphere, and by recognition of desirable map qualities, or map projection properties.

characteristics of the geographic grid

A listing of the characteristics of the geographic grid provides a ready check by means of which comparisons between the map projection and the spherical grid can be made. Where the projection displays a departure from the spherical grid, distortion of the map is indicated. Recognition of some principal kinds of departures and their related distortions is an aid in judging which parts of the map and which characteristics of the map are shown truly.

The check list given below includes items previously discussed in Chapter 2. It is emphasized that it has application only to the sphere.

1. The scale on the spherical grid is everywhere the same. Scale is a measure of a mathematical relationship between a distance on the globe, a projection, or a map, and the corresponding distance on the ground. Although it is constant on the globe it varies from place to place on *all* projections and on *all* maps. Since differences in scale affect measures of distance, they also affect measures of azimuths and the sizes and shapes of areas shown on maps.
2. All parallels are parallel and equally spaced along meridians. Where parallelism is not held on a projection the scale of the map developed upon it varies in a north-south direction.
3. Meridians are evenly spaced on the parallels and converge at the poles. Unequal spacing of equal measures of longitude along any parallel and lack of convergence indicate changes in the scale along east-west lines.
4. Meridians and parallels intersect at right angles. Where the right angle relationship is not held the shape of a small area centered on the intersection is distorted.
5. Between any two parallels of latitude, all areas having equal longitudinal measurements are equal in area. Where variations in the east-west scale are present on the projection, this quality is lost.
6. Any great circle, when viewed from directly above, shows as a straight line on the globe.

Analysis of a map projection in the manner suggested above indicates which of the map projection properties can be preserved in a projection and represented truly on the map. The properties are outlined below.

map projection properties

A map projection is said to possess a property when a particular quality in the map developed around the projection is shown truly. Properties may be grouped together as those which relate to areas and others which relate to lines on the map. In either case, however, variation in the *linear scale* (scale along lines) on the projection is involved. To describe variation, comparisons are made between the lengths of lines on the generating globe and the lengths of their representations on the projection. For example, where a line on a projection is shown equal in length to its corresponding line on the globe, it is said to be shown *truly*. As the standard for comparison, the scale of the line on the graticule is *true,* or *exact.* In contrast, if the line length on the projection is shorter than it is on the globe, its scale is *short,* or *small.* In like manner, where the line on the projection is longer than its global counterpart, the scale is *long,* or *large.* It should be recalled that the generating globe itself is the only true representation we have of the geographic grid.

An *equal-area* (equivalent) projection shows all areas on the map in their proper size relationships to one another. Thus, all one-inch squares on the map represent the same size areas on the ground, regardless of their positions. Since linear scales vary from point to point this may at first seem impossible. We note, however, that scale variations can be controlled. For any region on the map a too-large scale in one direction can be compensated for by a too-small scale in the other. Such modification produces the equivalent property.

In contrast, the *conformal* (orthomorphic) projection maintains the property of true shape in areas. In this projection the linear scale is held constant in all directions around a single point and the meridian and the parallel at the point intersect at right angles. In general scale varies from point to point. Consequently, shape can only be held true in small areas. In large areas its distortion may be extreme. Since the areal properties depend upon scale difference in the one, and a constant scale around a point in the other, it is obvious that no map can be both conformal and equal-area.

Other properties relate to lines. *Azimuthal* projections show the true azimuths of lines between one central point and all other points. In the case of the "other points," the azimuths of lines drawn between them are distorted. Similarly, an *equidistant* projection shows the correct distance between a central point and all other points. Distances will be distorted if they are taken between the "other points."

In contrast to the properties of true direction and distance, other line-related properties of importance are the straight line representations of great circle arcs and rhumb lines. Although on most projections these lines show as curves, certain projections, the gnomonic and the Mercator respectively, show them as straight and make them easy to work with. As noted previously, the great circle arc is the shortest line between two points on the earth's surface and the rhumb line describes a line of constant direction.

All map projection properties are not mutually exclusive as the equal-area and conformal properties are. Many can be combined with another. For example, the Mercator projection is both conformal and has a straight-line rhumb. Furthermore, the azimuthal property can be combined with either the conformal or the equal-area. Many other combinations are possible and lend to a projection the valuable two-property characteristic.

Characteristics of Some Projections

In the broadest sense, map projections may be classified in two general ways. What might be called a construction classification organizes them according to the means used in producing them; those that can be produced graphically, and others that are entirely mathematical. This is of principal use to the cartographer, whose task it is to make the map. On the other hand, the needs of the map-user are somewhat different. He must interpret the map. A projection classification wholly designed to suit his purposes might very well be confined to a consideration of the map projection properties. On this basis projections would be differentiated from one another in respect to those qualities which its map can or cannot show truly. In either case a knowledge of construction techniques as well as properties is useful. The classification used here is chosen largely because of its utility in describing both. Several of the more common projections are listed below and described on the basis of how they are produced. Certain details of construction which determine the arrangements of parallels and meridians are discussed in their relationships to the properties the projection possesses.

perspective projections

Graphic techniques produce the so-called perspective, or geometric projections. In these, points are transferred directly from the generating globe to the projection surface. The surface may be a plane (previously discussed) or one of the developable surfaces (a cone or a cylinder). Cones and cylinders are termed developable because, although they are curved, they can be flattened without distortions. If one or the other is placed tangent to the generating globe and light rays passed through the globe to it, a graticule is produced. When the surface is developed (flattened) it becomes a projection with a distinctive arrangement of parallels and meridians, and with distinct properties. For the conics and the cylindricals the light source is usually fixed at the center of the globe.

Perspective projections, then, differ with the projection surface and, in the case of the plane projections, with the location of the light source as well. They also differ in respect to their properties. These considerations form the basis for differentiating between them.

projections produced on planes: the planars

Figures 3.1, 3.2, 3.4, and 3.5 illustrate the principles of the plane projection. As noted, the locations of the light sources vary. The results of the variations show on the graticules and determine differences in the properties of each. In each case the projection plane is placed tangent to a point on the Equator to show the *equatorial aspect* of the projection.

In Figure 3.1, the *Orthographic Projection,* the light source is at infinity and all light rays passing through the generating globe are parallel. Lines of latitude are straight and parallel, and all meridians except the central one are arcs of ellipses. The spacings of equal longitude intervals along parallels are unequal, as are the latitude intervals

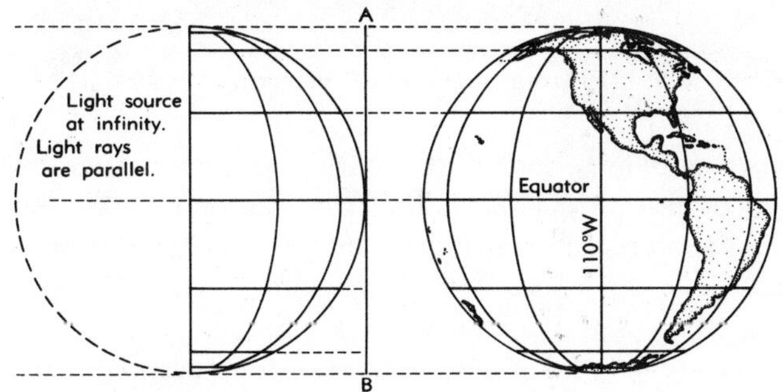

FIGURE 3.1. The Orthographic Projection

along the meridians. This means that scale varies both east-west and north-south. Since the variations are such that equal distances on the globe are shown to decrease outward from the center of the projection, the scale of the map representations of land and water areas *decreases* toward the periphery. This means further that the land masses shown are pushed to the outer edge of the map. They are compressed outward by linear scale distortions in all directions, and the relative sizes and the shapes of areas suffer as a result.

The orthographic projection is, in fact, a view of a hemisphere as we see it on a globe. Although it is not in common use it has been used to represent hemispheres. In recent years several fine atlases have used it, in modified form, to represent portions of the earth located at all latitudes.

In Figure 3.2, the *Stereographic Projection,* the light source is on the generating globe, antipodal to the point of tangency of the projection plane. In consequence, any light ray, except for the one which passes through the globe center, will pass through the globe at an angle. This has two effects. Any circle which is centered on the point of tangency will be projected as a circle, and all parallels and meridians other than the Equator and the central meridian will be projected as arcs of circles. Like the orthographic projection, meridians and parallels are unequally spaced along each other. Unlike the orthographic, however,

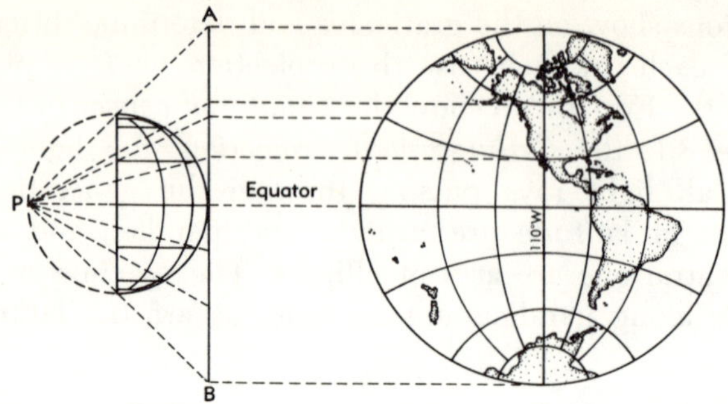

FIGURE 3.2.   The Stereographic Projection, Equatorial Aspect

scale *increases* outward, both north-south and east-west from the center. This has the double effect of pulling the mapped areas inward and of compressing their areas toward the center. In consequence, the relative sizes of areas are distorted, but in a manner which is quite different from the distortion noted in the orthographic.

Because meridians and parallels on the stereographic projection show as arcs of circles, the projection is easy to draw. Moreover, because they intersect at right angles, the maps drawn on them are conformal.

Although the stereographic is not in common use, it has been used for hemisphere maps, and when the central meridian has been properly chosen, it can be used for maps of South America, Africa and Australia. More specifically, because it is conformal and very nearly equal-area, the projection as viewed from the poles to show its *polar aspect* is used as a base for the Universal Polar Stereographic Grid (the UPS Grid System) on which all U.S. military maps of high latitudes are developed.

The stereographic projection is shown in its polar aspect in Figure 3.3. Here, the plane of projection has been shifted to a pole, and the light positioned at the opposite pole. Conformality is preserved, as is the nearly equal-area property. Moreover, direction from the point of tangency is shown truly. Like all projections, however, scale varies from place to place.

Because of its true shape property and the close approximations it gives to other desirable qualities in a map, the stereographic projection has been considered by some to be the best we have for showing the high latitudes.

Figure 3.4, the *Globular Projection,* is a compromise between the orthographic and the stereographic. As the notation on the figure shows, the light source is very specifically placed, outside of the generating globe. Both the lines of latitude and the lines of longitude are arcs of

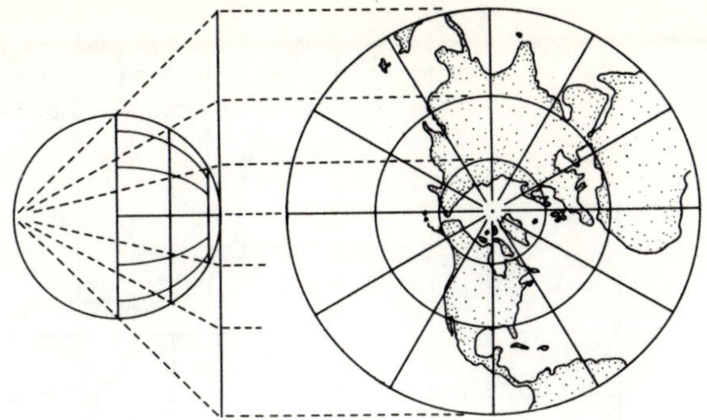

FIGURE 3.3. The Stereographic Projection, Polar Aspect

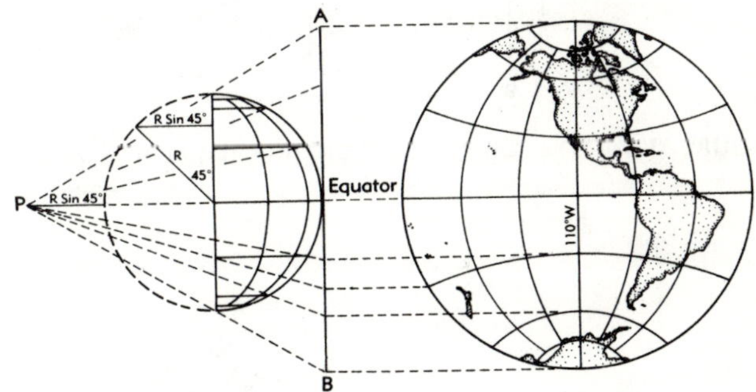

FIGURE 3.4. The Globular Projection

circles. Meridians are equally spaced throughout, and parallels are equally spaced at the periphery and along the central meridian. This arrangement relieves somewhat the scale variations shown on the orthographic and the stereographic. That such error is not completely removed, however, is evident from the fact that parallels are not shown parallel to one another.

The globular projection is neither equal-area nor conformal. It is, though, a close approximation to both and is useful as a school map to show hemispheres.

Figure 3.5, the *Gnomonic Projection,* is a wide departure from those projections previously discussed. The light source is at the center of the generating globe and, as a result, the poles cannot be shown. Meridians are straight and parallel, and the lines of latitude show as nonconcentric curves. All lines are unequally spaced, with scale increasing in all cases away from the center.

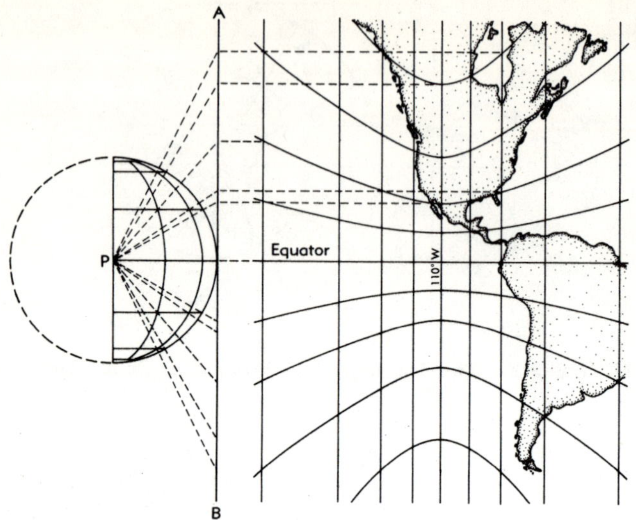

FIGURE 3.5. The Gnomonic Projection, Equatorial Aspect

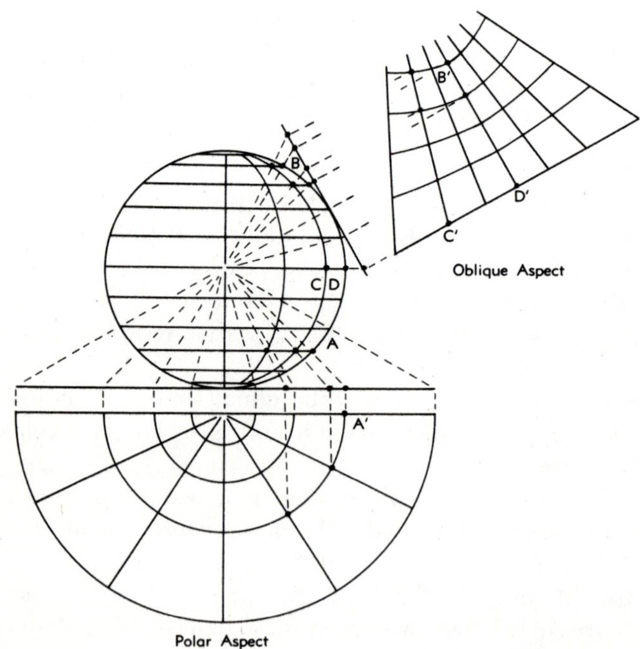

FIGURE 3.6. Gnomonic Projections, Oblique and Polar Aspects

The gnomonic projection has the useful property of showing all great circles as straight lines. Since great circle arcs are the shortest distances between points on the globe, and navigators are in part guided by them, their representations as straight lines on a map or chart are most useful. The projection, then, is chiefly used for navigational charts. The chief fault of the projection lies in the great distortions of lines and areas that become progressively greater toward the edges of the map and in the high latitudes.

The general weakness of the gnomonic projection in showing high latitudes is relieved somewhat by projecting it in either its *polar* or *oblique* aspect. As noted above, in the polar case the plane of projection is made tangent at a pole. In the oblique case it is made tangent at some mid-latitude. These positionings of the plane make it possible to chart the high latitudes. Since the shift of the plane has no effect on the properties of the projection, great circles are still shown as straight lines, despite the rearrangements of meridians and parallels which result from the shift.

Figure 3.6 shows both the polar and the oblique aspects of the gnomonic projection.

The straight line representation of the great circle arc may be thought of as the way the arc would appear to a viewer positioned at the center of the globe. From this point of view, any or all parts of the great circle would appear as a straight line drawn between two points on the globe surface.

projections produced on cones: the conics

As noted earlier, the conic projection is first produced on a cone held tangent to the generating globe with the light source positioned at the globe's center. The cone is then developed and a flat graticule produced. Although the projection is not used in precisely this form, many of our most useful projections are derived by modifications of it. In illustration of the principles involved, it is described below in a simplified form.

Figure 3.7 shows the construction of a *Simple Conic Projection.* The cone is drawn tangent to the globe along the *standard parallel*—in this case at latitude 30°. In the development the standard parallel is shown in its true length as the arc of a circle, with other parallels shown as concentric arcs of circles of different radii. The pole is the common center of the arcs. Meridians are shown as straight lines radiating from the pole. They are equally spaced along the parallels.

The dimensions of the development are derived in part by calculation and in part by graphic means. The radius of the standard parallel arc may be taken from the drawing, as the length of the line *AC*, or calculated as follows:

$$r = R \text{ Cot } 30°$$

Where: r = the radius of the arc
          R = the radius of the generating globe
          30° = the latitude of the standard parallel

Other arcs, corresponding to other parallels of latitude on the globe, have radii as indicated on the drawing, lines *AB*, *AD*, and the like.

On the development the length of the standard parallel is shown as an arc of a circle and is measured in *degrees* as:

$$360° \text{ Sin } 30° = 360° \ (.5000) = 180°$$

Where: 30° is the latitude of the standard parallel
          .5000 is the sine of the latitude angle and is
          referred to as the *constant of the cone*.

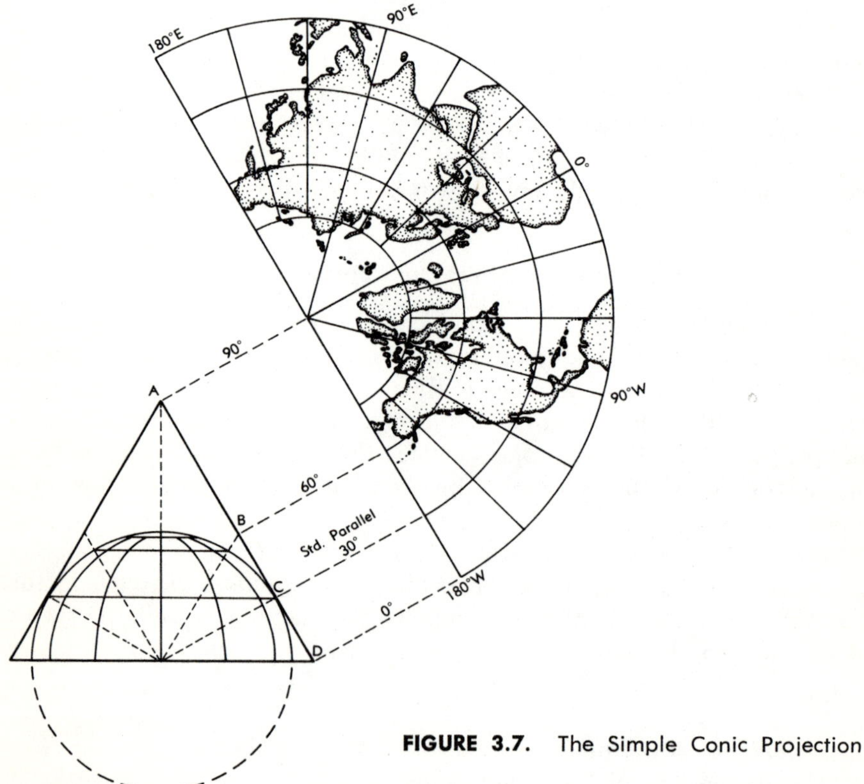

FIGURE 3.7.   The Simple Conic Projection

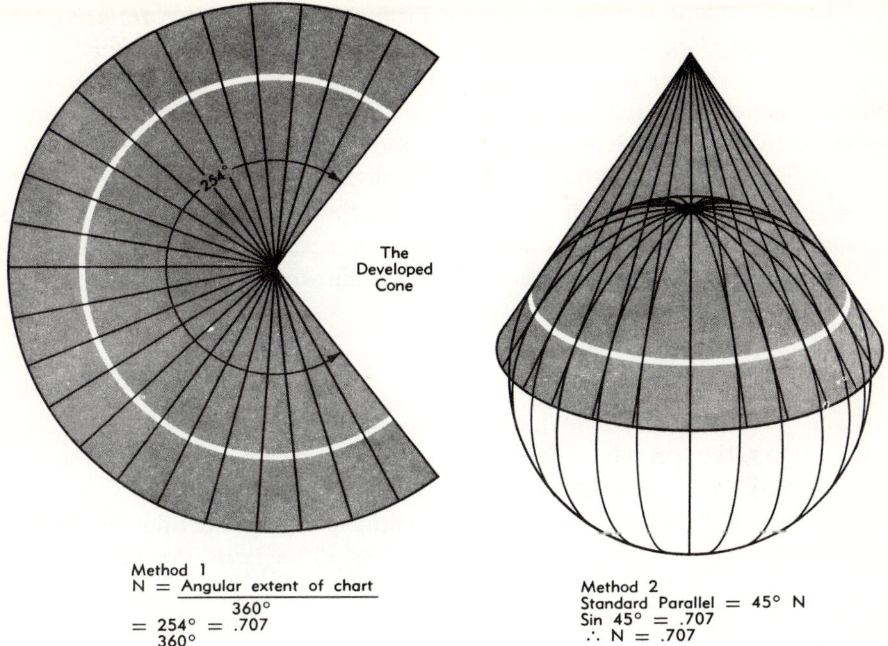

FIGURE 3.8. The Cone Constant (Courtesy of United States Air Force, from AFM 51-40, Vol. 1, Air Navigation.)

This means that all longitude angles are shown on the development at one-half their values on the generating globe. This relationship will, of course, change as the latitude of the standard parallel and the constant of the cone are changed. Figure 3.8 illustrates the principle using a cone tangent at 45° of latitude.

As described above, the projection is not satisfactory for a number of reasons, chiefly because scale distortions along the meridians become unmanageable at higher latitudes. It is improved by adjusting the distances between parallels. In this case, starting with the standard parallel, all parallels are spaced equally and truly along the meridians. When this adjustment is made it will be found that the pole is no longer shown as a point. It is, instead, shown as an arc of small radius. As a consequence of showing the dimensionless point as a line with a dimension, east-west distances at high latitudes are greatly distorted by being shown to be too long. Scale becomes too large, just as it is in latitudes which are lower than the latitude of the standard parallel. Scale is held true along the meridians and on the standard parallel.

The *Two Standard Parallel Conic Projection* (sometimes referred to as the *Secant Conic*) is a further improvement on the basic conic projection. It is shown in Figure 3.9. In this the cone is made to intersect the generating globe along two parallels. A parallel length is computed as the cosine of its latitude × the globe circumference. In the simplest case the spacing of the standard parallels on the projection is made equal to the length of the chord between the two on the globe. The spacing may be made true as is done on the simple conic, and thus locate on the projection two parallels of latitude shown in their true lengths and their true distances apart.

modified conics

The modified conics include several important projections. All are based upon the principles outlined above. Each is modified in some specific way or ways to give it a certain desirable property.

The principle of the secant conic is applied in constructing the *Lambert Conformal* and the *Albers Equal-Area Projections.* In Lambert's,

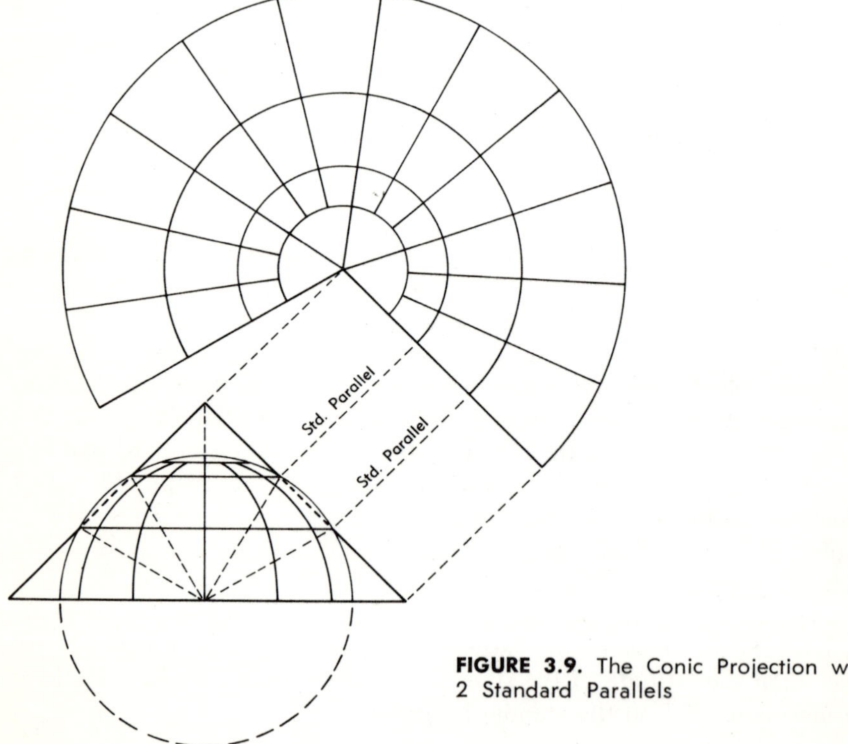

FIGURE 3.9. The Conic Projection with 2 Standard Parallels

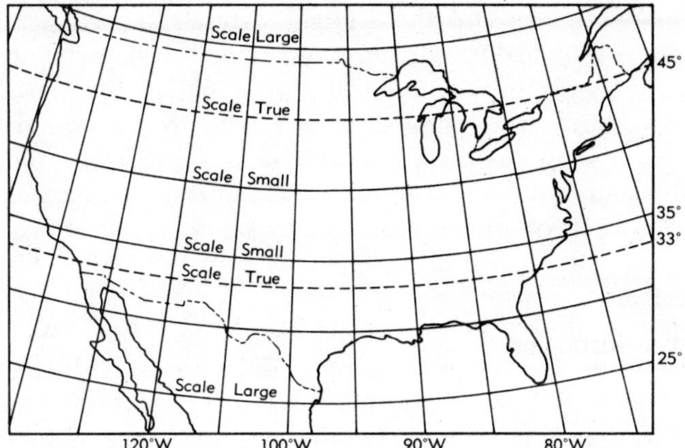

FIGURE 3.10. Lambert's Conformal Projection of the United States with 2 Standard Parallels at Latitude 45° N and 33° N.

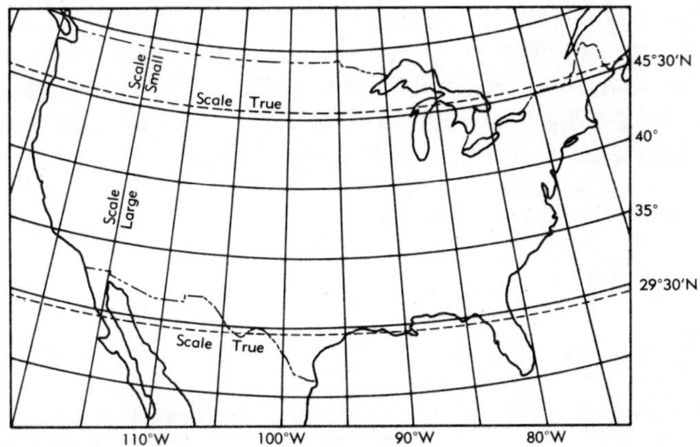

FIGURE 3.11. Albers Equal-Area Projection of the United States with 2 Standard Parallels at Latitude 45° 30' N and 29° 30' N.

the standard parallels are shown in their true lengths; others are not. Longitude intervals on the standard parallels are shown truly. On parallels between the standards, they are short; on other parallels outside the standards the intervals are shown too long. Scale varies accordingly. It is too small between the standards and too large beyond them. All parallels are so spaced along the meridians that, for every point, the scale on the parallel is the same as it is on the meridian. As in all conformal projections scale varies from point to point, but in the Lambert the variations are small. In maps of the United States, standard parallels have been chosen so that scale error does not exceed 2.25 per cent.

In the Albers Projection a similar principle is followed. Standard parallels are produced as before. All parallels are so spaced along the meridians as to show areas between them in their proper relative sizes. Between the standard parallels the scale along the meridians is too large, while beyond them meridian scale is too small. Greatest scale error in the projection, as it is used in making a map of the United States, is less than 2 per cent.

Both the Lambert Conformal and the Albers Equal-Area are particularly useful in showing areas in the mid-latitudes, with long east-west and relatively short north-south dimensions. They are not good for continents but are often used to show parts of continents, as in atlas maps. In the United States both are used in official maps, with the Lambert very often used in making aeronautical charts and serving as the base for state mapping. Albers is used for population density and other census mapping. Elsewhere, Albers has general purpose mapping utility, as in the U.S.S.R. and parts of Europe.

The *Polyconic Projection* is best visualized as a composite of sections of successive simple conic projections, with each of the cones drawn tangent to the generating globe at a different latitude. Its construction and its grid are shown in Figures 3.12 and 3.13. Parallels on the grid are

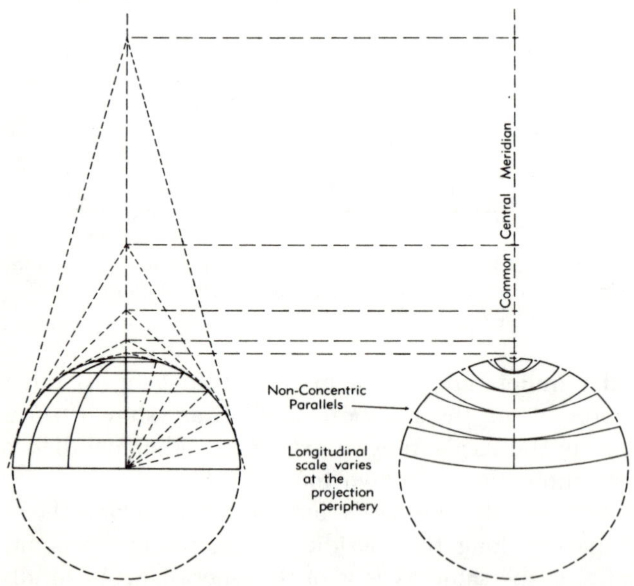

FIGURE 3.12. Construction of the Polyconic Projection

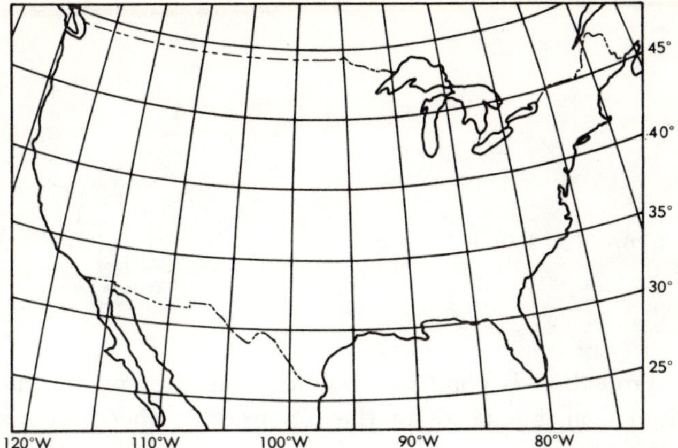

FIGURE 3.13.   Polyconic Projection of the United States

standard parallels of the cones. They show on the projection as non-concentric arcs of circles of different radii and are truly spaced along a common central meridian. The parallels themselves are truly divided for the meridians, which show as curves. Scale is held true along the parallels and the central meridian.

This projection is neither conformal nor is it equal-area. It is best thought of as a workable compromise. Its scale error is small near the central meridian but increases toward the periphery. For areas within 560 miles of the central meridian, error is less than 1 per cent. The maximum error for a map of the United States is 6 per cent.

Because scale is nearly true along the north-south central meridian, the polyconic is particularly useful in showing areas with long north-south and narrow east-west dimensions. It is used chiefly for topographic maps of relatively small regions, and was for many years the principal projection used by the U.S.C. & G.S. for maps of the United States.

The *Bonne Projection,* shown in Figure 3.14, is a one-standard parallel projection. In many respects its construction resembles that of the simple conic. Its standard parallel and central meridian are divided truly. Unlike the simple conic all of its parallels are also truly divided for the meridians. All parallels appear on the projection as concentric arcs of circles and all of its meridians, except the central one, show as curves which bend more sharply as they approach the pole.

The projection is equal-area, but scale error becomes excessive along the more peripheral meridians. Moreover, because of the scale error and the further fact that the angles between meridians and parallels depart considerably from a right angle toward the projection edge, the shape is increasingly poor away from the central meridian.

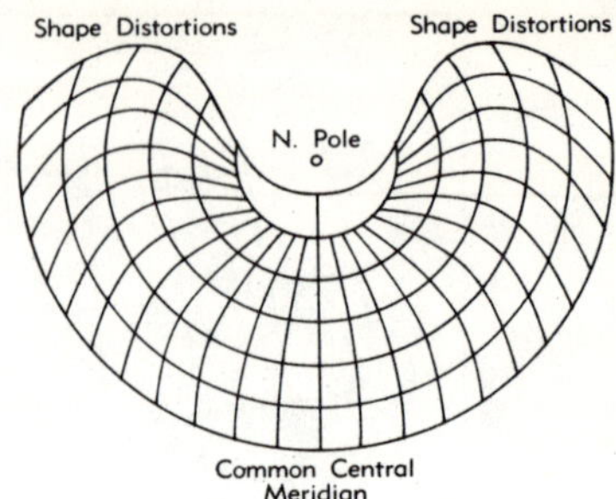

FIGURE 3.14. The Bonne Projection

Bonne's Projection is good for showing countries in the mid-latitudes and for this reason it is used for the topographic sheets of many of the nations of Western Europe. It has further utility in showing the continents of Europe, Asia, and North America.

projections on cylinders: the cylindricals

Cylindrical projections are illustrated by a cylinder held tangent to the equator of a generating globe. With a light source at the globe's center, points and lines on the globe are projected to an enveloping cylinder. When the cylinder is developed, a rectangular graticule is produced (Diagram A, Figure 3.15). The graticule is never used precisely as shown because of its numerous inadequacies. It does, however, provide a convenient beginning system of meridians and parallels which can be modified to produce other projections with useful properties.

As illustrated, the cylindrical graticule is made up of a number of straight meridians and parallels arranged to intersect one another at right angles. Meridians are equally spaced, but parallels are not and both are shown to be longer than they actually are on the globe. Only the Equator is shown in its actual global length and is represented truly. It is divided into equal longitude intervals which are carried into the higher latitudes by parallelism of the meridians. This stretches E-W distances, with the stretching increasing poleward; thus, the poles cannot be shown. Unequal stretching is also evident along the meridians and, consequently, equal increments of latitude on the globe are represented by lines of unequal length.

Distance distortions result in scale variations. Scale at the Equator is true. It increases elsewhere as latitude increases; at any latitude greater than zero the N-S scale and the E-W scale are different. Trigonometry tells us that the E-W scale measured along a parallel varies directly as the secant of the angle of latitude varies. In contrast, the N-S scale measured along a meridian varies somewhat more complexly as the tangent of the latitude angle varies.

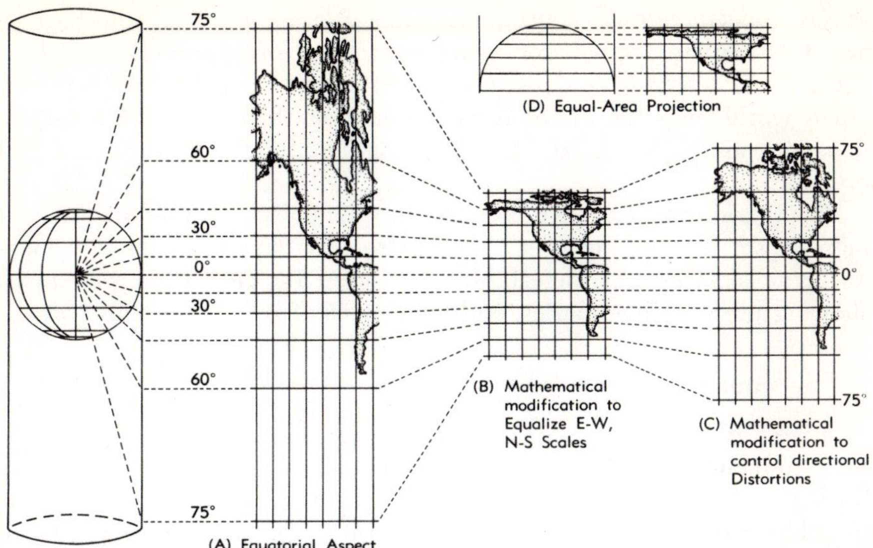

FIGURE 3.15. The Cylindrical Projection, Equatorial Aspect, and Related Projections.

Because of scale variations, the simple cylindrical projection is neither conformal, equal-area, equidistant, or azimuthal. Only the Equator is shown truly and the poles cannot be shown. All of its related projections illustrated in Figure 3.15 adjust some features of N-S scale and retain the true representation of the Equator so that areas or lines may be represented more realistically.

In Diagram D of Figure 3.15 the *Cylindrical Equal-Area Projection* modifies the N-S scales and achieves its property by changing the method of projection. Its graticule is produced by passing planes tangent to the poles and through the meridians and parallel circles of the generating globe. When extended the planes cut the cylinder. The graticule pattern is one of intersecting straight meridians and parallels, with the poles and the parallels shown as lines equal in length to the Equator. At latitudes greater than zero, N-S distances are shortened while E-W distances remain equal to those on the Equator. Scale variations are such that all areas are shown in their true relative sizes.

The Equal-Area Projection contrasts to the cylindrical in two significant ways: primarily, it has the equivalent property; secondarily, it can show the poles and the entire global surface. At high latitudes, however, distortions are too great for general purposes. The Equal-Area Projection can only be used for mapping in the equatorial and lower mid-latitude regions.

On the *Cylindrical Equi-Rectangular Projection* (Diagram B, Figure 3.15) equal degree arcs along meridians and parallels are given equal linear lengths. This produces a graticule of equally-sized squares. On a world map the effect is more pleasing than that of the projections previously discussed. However, high latitude distortions are still present.

This grid has considerable use in planimetric (two-dimensional) maps of small areas and is easy to construct; if the area is small enough, no great scale error is detectable. It is used for city maps, plat maps showing property boundaries, and road maps of relatively small areas.

mercator projections

Like the foregoing, the *Mercator Projections* employ the principle of introducing a modification into the cylindrical. Mathematical operations are used to equalize N-S and E-W scales around individual points to achieve the conformal property. Mercator Projections are produced in two basic forms, each of which has specific use, in either navigation or world mapping. They are illustrated in Figures 3.15 through 3.17.

In its *conventional* form, the Mercator is a modification of the cylindrical projection in the equatorial aspect. The general principle applied in its construction is illustrated by comparing Diagrams A and C in

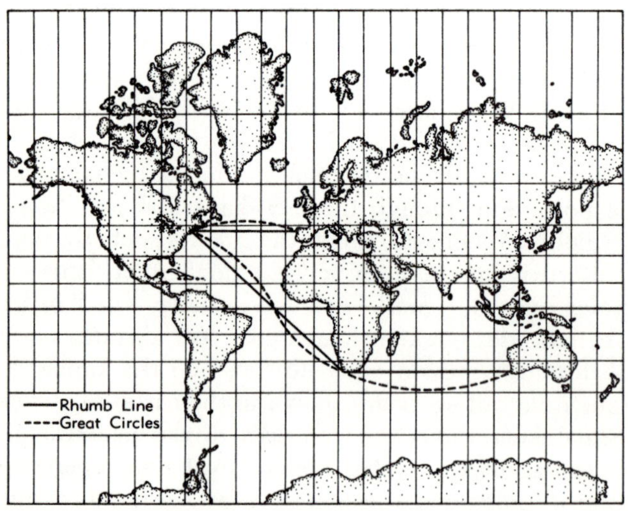

FIGURE 3.16. Map of the World on the Mercator Projection

Figure 3.15. Primarily we see that on the Mercator (Diagram C) longitude intervals remain as they are on the cylindrical. Parallels, however, have been moved toward the Equator. This means that the equatorial scale of the cylindrical has been retained on the Mercator, while all N-S scales have been decreased. In practice they have been decreased to the extent that on the Mercator, at any latitude, the N-S scale is equal to the E-W scale.

The mathematical principles involved in making the necessary N-S scale adjustments along the meridians are illustrated by reference to the 60° parallel. On the generating globe the length of the parallel is cosine 60° or 0.5000 times the length of the Equator. On the cylindrical projection the two are shown to be equal. This means that on the projection the E-W *scale* along the parallel is twice the *scale* along the Equator. In other words, the *scale* along the parallel equals the reciprocal of the cosine (the secant) of its latitude times the equatorial scale. This general rule is applied in spacing all parallels on the *Conventional Mercator* grid. On the grid, small increments of meridians are drawn at scales equal to the scales of adjacent parallels.

The Conventional Mercator Projection has its best use in navigational work, because, in addition to its conformality, it has the straight rhumb line property. Used in conjunction with the great circle gnomonic charts, the straight rhumb line is invaluable to the navigator. Also, because it is conformal and gives good approximations of areas and the lengths of lines in low latitudes, the Conventional Mercator can be used in these regions for charting short stretches of coastline and harbor areas and to show areal distributions on larger maps. On the negative side, this projection cannot show the poles and the sizes of areas in the high latitudes are grossly enlarged. Rarely is it used for mapping in areas located at latitudes greater than eighty degrees.

The basic weakness of the Conventional Mercator is corrected on the *Transverse Mercator Projection,* which permits conformal mapping at the high latitudes. This projection adjusts E-W scales and makes them equal to the N-S scales that would appear on a cylindrical projection produced in the meridional aspect. In this aspect, the cylinder is rotated 90° and made tangent to a meridian circle. Figure 3.17 shows the projection after adjustments have been made and compares the Transverse Mercator grid to the grid of the conventional. As shown in the figure, the central meridian is a representation of the tangent circle and is produced in its true length, at exact scale. Latitude intervals along it are also shown truly, as they were on the generating globe and the original cylindrical projection. Longitude intervals, however, have

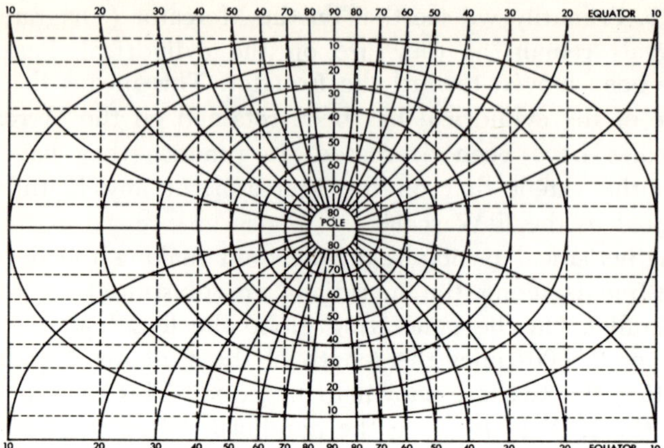

FIGURE 3.17. The Transverse Mercator Grid with the conventional grid superimposed. (Courtesy of United States Air Force, from AFM 51-40, Vol. 1, Air Navigation.)

been decreased to make the E-W scales equal to the N-S scales at individual points. The principle followed is similar to that used on the conventional grid. For example, on the Equator as shown in Figure 3.17, at longitude 30°, the meridian is 60° away from the central meridian. Scale along the Equator at this point is calculated as secant 60° times the scale along the central meridian. East and west of the central meridian the Equator is a line of constantly increasing scale and a full 180° of longitude along it cannot be shown.

A refinement of the Transverse Mercator applies the Mercator principles to a secant cylinder, a concept not unlike that of the secant cone. The diameter of the enveloping cylinder is smaller than that of its generating globe and the cylinder is caused to intersect the globe along two small circles equidistant from the central meridian. On the projection the small circles show as straight lines, parallel to and equidistant from the central meridian and drawn to exact scale. They enclose a narrow longitudinal zone within which all scales are small, while outside the zone all scales are large. In effect, this procedure increases the E-W scales away from the central meridian in much the same manner as does the Transverse Mercator which is derived from a conception of a tangent cylinder in the meridional aspect.

In its transverse forms the Mercator retains its conformality but loses its straight rhumb line. The Transverse Mercator in its secant form has its best use as the basis for conformal mapping in narrow zones of longitude, particularly in the U. S. State coordinate systems and in the Universal Transverse Mercator (UTM) Grid System. It serves well, first, for states with long N-S and narrow E-W dimensions; and second, for mapping the world in narrow zones. Each of the zones is mapped on its own central meridian and has its own set of lines of exact scale. The related grid coordinate systems are discussed in greater detail in Chapter 4.

the mathematical projections

Unlike the perspectives, mathematical projections are completely mathematical in their derivations. They are developed with no consideration being given to the projection surface or the light source. Each is calculated for a specific purpose and most can be used to show one or two characteristics of the earth grid truly. Some are designed as composites and show all characterististics of lines and areas *almost* correctly. Some that are in common use are discussed below.

Figures 3.18 and 3.19 show the polar aspects of the *Azimuthal Equidistant* and the *Azimuthal Equal-Area Projections.* As their names imply, both show azimuths correctly from a central point. Each shows its parallels as concentric circles and its meridians as straight lines radiating outward from the common center. In the equidistant, parallel circles are equally spaced along the meridians to show distances from the center truly. For the equal-area, parallels are so spaced that areas between them are shown in their true size relationships to one another. Each is, then, a two-property projection. For each its two properties will carry over into either the equatorial or the oblique case. Obviously, these projections are highly useful. The equidistant is used for airline maps and radio and seismic work, to trace air routes, radio transmissions, and earthquake shocks from central points. On the other hand, the equal-area projection has common use in showing hemispheres.

On the *Sinusoidal Projection* the Equator and all parallels are laid out in their true lengths and divided truly. The central meridian is perpendicular to the Equator. It is shown in its true length and also divided

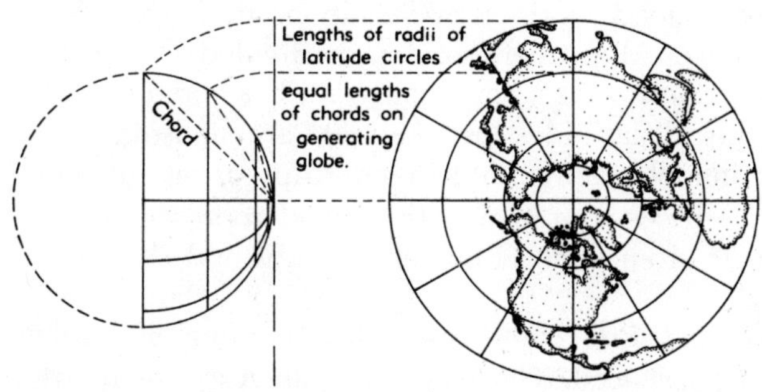

FIGURE 3.18. The Azimuthal Equal-Area Projection

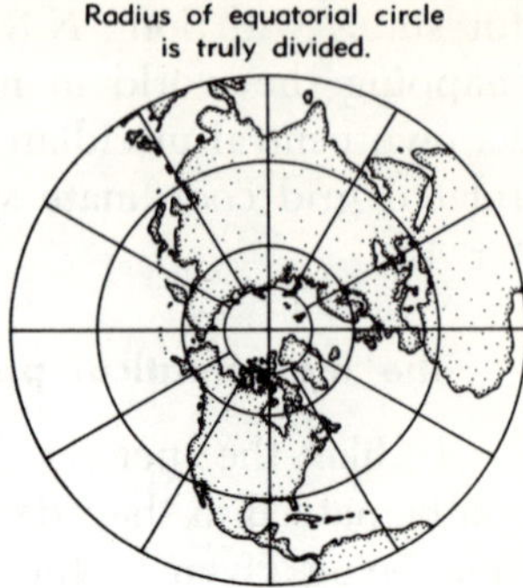

FIGURE 3.19. The Azimuthal Equidistant Projection

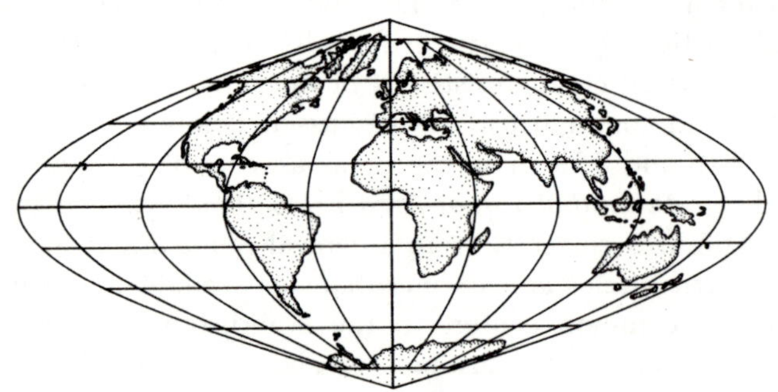

FIGURE 3.20. The Sinusoidal Projection

truly. Thus all parallels are held parallel and are equally divided. The meridians drawn to intersect them are sine curves.

The sinusoidal is equal-area but, because of its high latitude east-west distortions, its use is limited to low latitude regions. It is sometimes used on maps of Africa, South America, and Australia.

Mollweide's Homolographic Projection is an oval, with a central meridian drawn perpendicular to the Equator, at half the equatorial length. Meridians are ellipses and divide the parallels truly. Parallels are straight and parallel and so spaced along the central meridian as to make the projection equal-area. When centered on Europe, as shown in the diagram, it gives a good world representation. If, however, the central meridian is placed on the Americas, Asia is divided.

Goode's Homolosine Projection is a combination of the homolographic and the sinusoidal. In this all latitudes less than 40° are sinusoidal, and all latitudes above are homolographic. When interrupted, or cut along selected meridians, it combines true area with a good representation of shape for a world map. It often appears on school maps and on illustrations in texts.

Denoyer's Semi-Elliptical Projection resembles the homolographic, but its parallels are equally spaced. The poles are shown as one-quarter

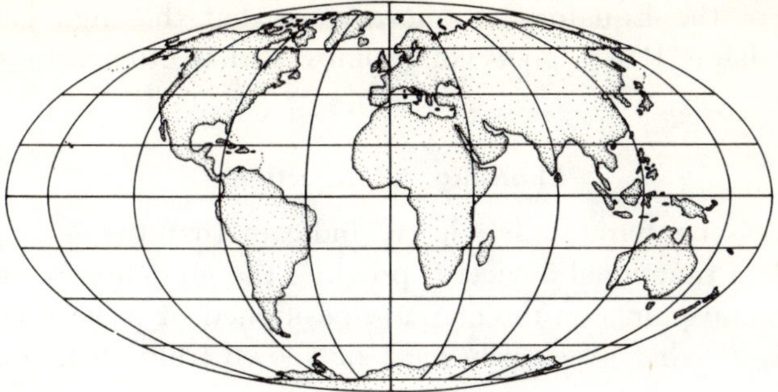

FIGURE 3.21. Mollweide's Homolographic Projection

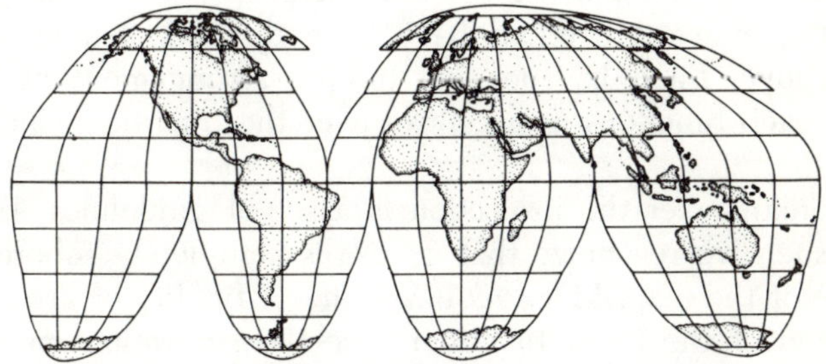

FIGURE 3.22. Goode's Homolosine Projection

(Copyright by the U. of Chicago, Goode's Base Map Series, Dept. of Geography, University of Chicago.)

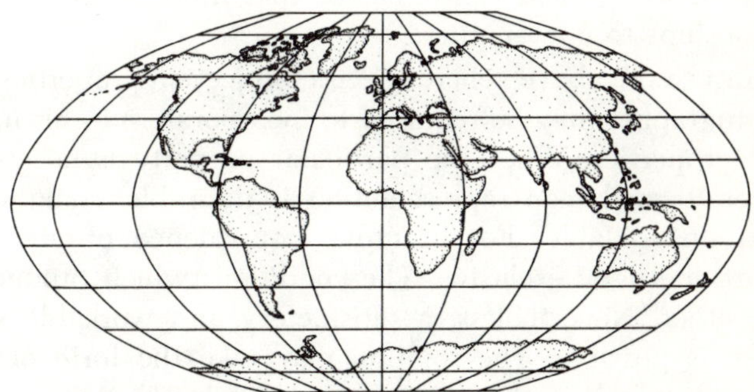

FIGURE 3.23. Denoyer's Semi-Elliptical Projection

Copyright by Denoyer-Geppert Company, Chicago (Used by permission.)

the length of the Equator, to relieve somewhat the high latitude distortions of shape. It is a good compromise and is often used on schoolroom maps.

Choosing A Projection

The foregoing paragraphs clearly indicate that the map projection is, above all, a functional device. It provides the guidelines around which areas, lines, and points on the map are positioned. A projection is chosen on the basis of what necessarily must be shown truly on its map. As we have seen, it is a mathematical fact that the flat map cannot possibly show all characteristics of earth surface features with complete accuracy. We have also seen that by a proper choice of the projection, some characteristics of that surface can be shown on the map as they truly are, at reduced scale, of course. As to which characteristics can be shown truly, this depends entirely upon the property or properties of the projection. Consequently, many projections have been devised. A few have been described above. Each has its own unique arrangement of lines and points, and each functions as a control over the qualities which its map possesses.

For the mapmaker the projection is a set of guidelines. When it is viewed as such we recognize that its use is completely subordinated to the purpose of the map. Maps require either a fidelity of area, or a true representation of the lines. In either case the projection must suit the purpose of the map or the map loses its effectiveness as a means of presenting factual information. Thus a map used to demonstrate areal variations in population density is drawn on an equal-area projection simply because a portrayal of density differences is a comparison of areas of equal size. In contrast, a map used to illustrate long distances is based on a projection which minimizes scale error. Other purposes may require that areas be held conformal or that lines be oriented truly in their relationships to one another.

In contrast to the above, where specific projection properties are called for, the cartographer may be required to incorporate into his map a compromise of properties. Where no particular property is necessary but a general overall good map representation is desirable, some sacrifice of accuracy is unavoidable. For example, equivalence of area and conformality are mutually exclusive. The conditions which produce the one negate the other. Nevertheless, a satisfactory and workable solution to the problem is possible. The solution may take the form of an equal-area projection which is almost conformal (as in the Albers) or it may be almost conformal and almost equivalent (as in the globular). In

TABLE 3.1

Conformal Projections

Projections	Uses
Stereographic	Hemispheres, Polar Regions
Lambert's Conic	Aeronautical Charts, areas in mid-latitudes, with long E-W dimensions
Conventional Mercator	Navigational Charts, maps of equatorial regions
Transverse Mercator	High latitudes, areas with long N-S dimensions

either case the purpose of the map is fulfilled and a projection which displays no gross distortions is available to suit the conditions.

Attention to the properties of projections is not enough by itself. Consideration must also be given to the possibility of creating a simplicity or a generalized realism in the map. Great distortions of size and shape should be avoided where possible. A map of a large area, for example, a continent can be made on the conventional Mercator. In this case conformality or the straight rhumb may be interesting but not necessary. However, because of size distortions in the high latitudes, the Mercator-based map would lose much of its effectiveness when no more than a general overview is required. A stereographic or even a globular would perhaps serve as well and add the virtue of greater realism to the whole.

As a final point which bears upon the problem of selecting a projection, consideration should also be given to the size, shape, and location of the mapped area. Quite clearly, for large scale maps of relatively small areas, distortions of any kind are minimized. Consequently, the equi-rectangular grid is adequate for many purposes such as showing city plans, generalized land uses within the city and the like. On the other hand, small scale maps of the world or a hemisphere might very well require a distinct property. For these there is always the very real problem of achieving the required property without creating some gross distortions of other kinds. To solve this difficulty, the stereographic, the homolographic, and the interrupted homolosine have much to recommend them.

Maps of areas with long east-west dimensions require minimal distortions along their parallels. For these the modified conics serve well in the mid-latitudes, as the conventional Mercator and other mathematical projections do for the equatorial regions. On the other hand, for areas which are elongated in the north-south directions, both the transversed Mercator and the polyconic provide a basis for good representations along the central meridian.

The above considerations are presented in a more systematic fashion in the adjoining tables. Some of the more commonly used projections are grouped together in accordance with their properties and their principal uses noted.

TABLE 3.2

Equal-Area Projections

Projections	Uses
Albers	Maps of the U.S. and other mid-latitudes
Azimuthal Equal-Area	Hemispheres and continents
Bonne's	Mid-latitudes
Homolographic	World maps and hemispheres
Sinusoidal	Equatorial areas

TABLE 3.3

General Purpose Projections

Projections	Uses
Polyconic	Topographic sheets of the U.S., mid-latitude areas with long N-S dimensions
Equi-rectangular	Maps of small areas, city maps
Globular	Hemispheres

TABLE 3.4

Special Purpose Projections

Projections	Uses
Conventional Mercator	Rhumb lines are shown as straight lines, navigational use
Gnomonic	Great circles are shown as straight lines, navigational use
Azimuthals	Straight lines from the center are lines of true direction. The projection may also be made equidistant or equal-area.

Map Elements

A map is a graphical representation of a part or all of the earth's surface produced on a plane. It is drawn to some scale and the information it contains is plotted on a grid in a symbolized form. As a device which shows the nature and the relative locations of earth surface phenomena, it is an indispensable tool for geography and an aid to many other kinds of study as well. Like other tools of expression, however, it has both its limitations and its virtues, and these must be understood if it is to be used effectively.

The Utility of Maps

Maps are prepared to serve a variety of purposes. Wherever and whenever the locations or the geographical distributions of phenomena are a concern, the use of the map is indicated. Such occasions often arise in business and government operations where the map is used in such varied activities as regional planning, census-taking, military undertakings, and marketing and transportation analyses. In the academic fields maps may also be used in studies in a number of fields other than geography itself. Oceanographers, geologists, and climatologists all have interests in the geographical arrangements of the physical phenomena they study, just as the historian and the economist concern themselves with the locations of man and his activities. For these specialists, and for their governmental and business colleagues too, the map is a useful tool. It serves as a record of observations and a source of information concerning the "what" and "where" of features which characterize the earth's surface. Like any record or any source, the map is a depository of factual information which serves to communicate ideas and to stimulate the imaginations of workers in many fields.

Because maps range so widely in their purposes, they vary considerably in respect to the kinds of information they convey. Nevertheless, all maps share certain common qualities. They are, first of all, generalizations about specific parts of the earth. All are drawn to scale and are smaller than the areas they represent. As a result, they cannot possibly show all of the detail of the reality of the portion of the earth's surface they are meant to cover. They can, however, show some selected aspects of this reality by the use of symbols.

Scale, selective symbols, and locational systems are the elements from which all maps are developed. They are common to all true maps and are defined on maps either in a legend or by marginal notations. For special maps the elements are made to suit the special purpose of the map. When the map is constructed for more general purposes, the elements are standardized wherever possible.

Map Scale

Scale is a statement of the ratio between the size of a map representation of an object and the actual size of the object on the ground. It may be expressed mathematically as the so-called representative fraction (R.F.), graphically (in a diagram), or verbally.

The representative fraction is the reciprocal of a number in which unity (the numerator) refers to a map measurement, and the denominator is the corresponding measurement taken on the ground. A fractional scale of 1/1000, or 1:1000, means simply that one unit of measure on the map represents 1000 units of the *same measure* on the ground. Thus a 1-inch map distance corresponds to a 1000-inch ground distance, a 1-foot map distance corresponds to a 1000-foot ground distance, and so on. The scale of a map is larger than that of another if the denominator of its representative fraction is smaller. For example, a 1/1000 scale is larger than a 1/10,000 scale, because the representative fraction is numerically larger.

linear and areal scales

All maps have two kinds of scale: *linear* and *areal*. Linear scale refers to lines, and areal scale to areas. As pointed out in Chapter 3, the linear scale varies and can only be held constant along a particular line or along lines radiating from a point. Because of variation, it is convenient to designate a map scale in terms of the *nominal* linear scale. The nominal scale, as given in the map legend, is usually most nearly approximated in areas near the map center. If the nominal scale is the same as that of the generating globe from which the projection has been taken, it may be referred to as the *natural* scale.

In contrast to linear scale, the areal scale can be held constant all over a projection and its map. For equal area projections and maps, it is always equal to the square of the nominal scale. Thus, a 1:1000 map has an areal scale of 1:1,000,000. One square inch on the map represents 1,000,000 square inches on the ground.

For true maps, some notation of nominal linear scale is always given in the legend. Scale may be stated simply as a representative fraction, or it may be shown as a small bar scale (See Figure 4.1). For some equivalent area maps, an areal scale may also be shown. This too can be expressed as a fraction, or shown in a diagram as a rectangle which is meant to represent a stated size area on the ground. Areal scales occasionally appear on United States Census maps based on the Albers Equal-Area Projection.

From a theoretical standpoint, maps can be drawn at any scale desired. In practice, most maps are drawn at linear scales which are even multiples of 1/100, as 1/1000, 1/25,000, 1/1,000,000 and so on, as a concession to the metric system. This follows from the fact that the geodetic surveys upon which the maps are based are related to the ellipsoid which is first defined in metric measurement terms. In countries where the English system of land measurement is in common use (United States, United Kingdom and Commonwealth Nations) graphic scales which relate yards and miles to meters and kilometers usually appear in the map legend so that quick conversions can be made. General practice may also state scale as an approximation. A scale of 1:62,500 is often described as an "inch to a mile" scale (verbal). The proportion 1:62,500 closely approximates the 1:63,360 scale which is based on English system measures and is a true "inch to a mile" proportion (12 inches × 5,280 feet = 63,360 inches per mile).

scale and map accuracy

Quite clearly, the scale of a map affects its accuracy. On a 1:1,000,000 map, 1/16 inch represents approximately 1 mile; on a 1:250,000 map, 1/4 inch represents the same. It is apparent that on these maps it is impossible to obtain the same degree of precision as is possible on maps of larger scale such as 1:62,500. For example, on most government maps of continental United States, 90 per cent of all points shown are located within 1/50 of an inch of their true map positions. This means that on the 1:1,000,000 map error is held to about 1/3 mile, on the 1:250,000 map it is held to about 1/12 mile, and on the 1:62,500 map it is held to 1/50 of a mile. At other scales, point locations are held to corresponding degrees of precision.

A further consideration of accuracy relates to the thickness of lines on maps. Although the normal human eye can distinguish much finer lines, as a general rule the finest line used on a map is about 1/250 of an inch wide. On the 1:1,000,000 map this corresponds to about 330 feet on the ground, while on the 1:62,500 map it represents slightly more than 20 feet. In reality then, lines on maps represent zones on the ground, and the widths of the zones will vary with the map scale.

scale categories

Because of the scale-accuracy relationships, common cartographic practice recognizes three general scale categories of maps. Maps are classified as small, medium, and large scale according to their nominal linear scales and the amount of detail they can show. *Small scale* maps of 1:1,000,000 and less are highly generalized and used to represent large areas. If the scale is 1:5,000,000 or smaller, it is sometimes referred to as *global.* These maps can do little more than show the most gross of distributions, the broad outlines of major features, and the generalized orientations of large rivers and routeways. They contrast to *large scale* maps drawn to scales of 1:250,000 and larger. Large scales are used for topographic sheets, the basic maps published by governmental surveying agencies, for cadastral (real estate) maps, city and county maps, and the like. They can show considerable detail such as minor land and water features, the outlines and details of cities and villages, all kinds of roads including trails, and such relatively small features as individual buildings. *Medium scale* maps, as their name implies, are drawn to in-between scales which range from 1:1,000,000 to 1:250,000. Like the smaller scaled maps, they are highly generalized, but permit more detail in the configurations of the outlines of areas. They are often used in atlases, and for road and airway representations.

changing scales

In the construction of medium and small scale maps basic data are compiled from maps of larger scale which are based on ground or air surveys. Quite commonly the map compiler is faced with problems of changing scales. Scale changes may be made by graphical, photostatic or photographic methods, or by uncomplicated mathematical procedures. If, for example, it is required to change a 10 inch distance on a 1:50,000 map to its proper proportional length on a map of 1:1,000,000 scale, it is a simple matter to divide 1/1,000,000 by 1/50,000 and derive an answer of one-twentieth. The distance on the smaller scale map will then be 1/20 of 10, or 1/2 inch. The rule is: *divide the smaller scale by the larger and multiply the distance which is to be reproduced by the result.*

map scale and symbolism

On the ideal map every feature on the mapped portion of the earth would be shown in its true shape, orientation, and proportion. This is of course impossible, not only because of the errors which are inherent in the map projection, but for reasons related to the scale of the map as well. This is evident when we consider that on a map of 1:50,000, a square mile would show as a small square about 1 1/4 inches on a side. Now if every feature were plotted true to scale on such a map the results would be impossible to read even if a magnifying glass were to be used. For a map to be readable symbols must be employed. Many must of necessity be exaggerated in size. For example, at the 1:50,000 scale the standard symbol for a single track railroad covers a line width which is equivalent to about 165 feet on the ground, while a small house symbol covers an area on the map which represents a square approximately 85 feet on a side on the ground.

Symbolizing Maps

Map symbols are distinctive markings on maps which represent earth surface features or conditions as they are perceived on the ground. The feature mapped may be an object with distinctive form and dimensions, or it may be an area which is defined according to some outstanding characteristic. Symbols may be points, lines, two-dimensioned figures, or patches of color or pattern. In any case, the symbol has the purpose of giving identity to some portion of the map.

symbolization of areas

Areas are symbolized on special purpose areal maps, most of which fall into one of three basic map categories. One category divides the land into areas or regions in accordance with some classification system. Soil maps, for example, show areas where certain kinds of soils are dominant. The political map shows states, counties, and other political units as areas; and the vegetation map may show areas of forest and grassland. These maps and many others like them show differences in the kinds of areas. Areal boundaries represent abrupt changes in some quality which characterizes the earth's surface. Areas are symbolized and differentiated from one another by patches of color or pattern, and the map thus developed is commonly referred to as color-patch, or *chorochromatic* (from the Greek, *choros,* meaning a place, and *chromatikos,* meaning suited for color). The map of time zones shown in Chapter 8 is of this type.

In contrast to the above, a second category of map may be drawn to show variations in the intensity of a single phenomenon which has a general distribution over the earth. Variations are shown by use of *isolines* (from the Greek, *isos,* meaning equal), or lines which connect points of equal value. Temperature and rainfall maps, contour maps which show elevations, and air pressure maps are of this type. For a few of the many kinds of isolines in common use, a terminology listing is given below.

Isohyet	— equal precipitation
Isotherm	— equal temperature
Isohypse	— equal elevation
Isobar	— equal pressure
Isobath	— equal depth (of water)
Isogonic	— equal magnetic variation of the compass
Isohaline	— equal salinity in the oceans
Isochrone	— equal time of travel

Isoline maps have the great advantages of indicating transitional changes between areas. They are exemplified in the map of isogonic lines shown in Chapter 2, and in the common topographic map which shows the elevations and relief of the earth's surface. Topographic maps are discussed in a later section under the general heading of *the representation of relief.*

The third category of special purpose map is the density map. It may make use of any one of three basic methods of symbolizing areas. Density is a mathematical statement of the number of items found in a given area, such as 500 persons per square mile or 1 acre of arable land per 100 acres. When calculated for a statistical area it may be shown by color-patch techniques, or by dots. Because density relates to area, the color-patch is the most valid method of showing it, albeit somewhat crudely and generally. The color-patch covers the statistical area on the map and serves to differentiate it from other statistical areas on the basis of numerical data. This, however, assumes that the item of concern is evenly distributed within the statistical area. It may, then, be judged appropriate to take the total number of items within the area, represent this number by a dot, and place the dot at the center of the statistical area. This gives us the dot map when all areas for which statistics are available are symbolized in this way.

It may also be judged appropriate to show densities by means of isolines. In this case, dots which have equal values are connected by the lines. Other isolines are interpolated between those which have calculated values. Like other isoline maps, this map has the virtue of indicating transitional changes within areas. It must be pointed out, however, that the construction of either a dot or an isoline map requires considerable judgement and knowledge of the area mapped. Each of these two kinds of maps depends upon pinpointing the locations of many points to which values have been assigned. Since the numerical data used tend to project a high degree of certainty to the mind of the map-user, the mapmaker has the obligation of locating point values with a comparable degree of accuracy. This kind of accuracy requires the cartographer to know the mapped area and to know it well.

the representation of relief

The relief of an area is the difference between the highest and the lowest elevations in the area. In contrast to elevation, which is a point feature, relief is a feature of areas and is symbolized by areal symbolization methods. Relief and its associated point elevations are shown on *topographic* maps as the third dimension of the earth's surface (Figure 4.1). In this sense, topographic maps are differentiated from *planimetric* maps which show only two dimensions. The topographic map shows the topography of an area by using one or more of several methods of representing relief features.

The relief of an area may be shown by superimposing spot (point) elevations on a planimetric map. This representation may be improved by drawing *contour lines* connecting the points where elevations above some common datum (usually mean sea level) are equal. The addition of the lines provides a graphical representation within which the forms of relief features are revealed. Where adequate point data are not available, *form lines* may be used to give approximations. Contour lines are normally printed in brown but may be shown in gray on medium-scale maps. On medium and small-scale maps contours may be supplemented by hill-shading or the addition of color bands or layers between selected contour levels.

Contour lines are drawn on maps at definite elevation intervals. They are illustrated in Figure 4.1. The *contour interval* is the *vertical* distance between two contour lines. If the interval is held, the contour lines on the map are shown to be far apart in flat areas, but close together where

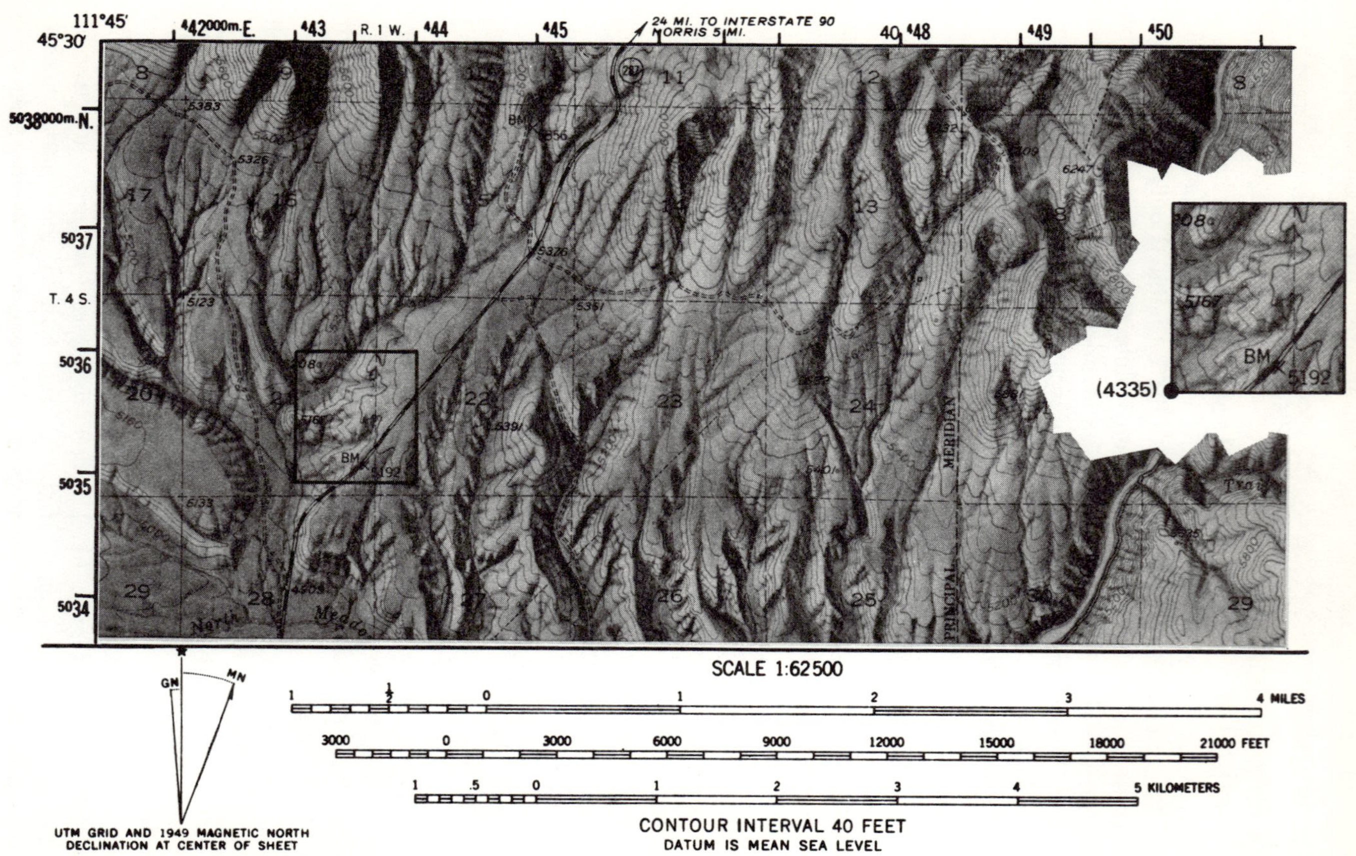

FIGURE 4.1. Portion of U.S.G.S. Topographic map, Ennis, Montana

slopes are steep. The interval is chosen to suit the nature of the terrain. Intervals are small where slopes are gentle, and greater on representations of hilly or mountainous lands. As an aid to the map-user, every fifth contour line (the *index contour*) is drawn heavier than the others.

Hachures may also be used to show relief features, either alone, or as supplements to contour lines. They are short lines which indicate clearly the direction and relative degree of slope. Steep slopes are shown by short lines drawn close together, while gentle slopes are shown by longer and thinner lines. Hachures often appear on European maps but are little used on maps produced in the United States.

object symbols

A map may be designed for the specific purpose of showing unusual objects and for this reason require the use of special symbols. Most maps, however, show some objects which recur from place to place over the earth's surface. These more or less common objects are shown on maps by standard topographic symbols, or conventional signs.

To the extent that it is practicable, an object is shown by the same symbol on maps of different scales, although circumstance may require that the symbol be modified somewhat. As a rule a symbol resembles the object it represents. When positioned on a map its center and its orientation are made to correspond with the true center and orientation of its object. In the case of linear features such as roads and streams, the symbol is made to retain the actual variations of alignment as they are observed on the ground. Where the symbol is shown at an exaggerated scale, other nearby symbols may be displaced.

Standard topographic symbols are usually divided into five categories which appear on maps in color.

1. *Cultural,* or man-made features are shown in black. Some roads are shown in red. On aeronautical charts they may be shown in brown.
2. *Hydrographic,* or water features such as lakes and streams are shown in blue.
3. *Hypsographic,* or relief features are shown in brown or gray.
4. *Vegetation* and *cultivation* features are shown in patterns of green.
5. *Special* features, such as those used on aeronautical charts may be purple, yellow or orange.

Some representative symbols are shown in Figure 4.2.

Supplementary Coordinate Systems

As pointed out in previous chapters the basic locational system, the geographic grid, is structured around a conception of angular coordinates and curved lines. As it was further pointed out, it is oftentimes convenient to superimpose rectangular, straight-line grids based on linear coordinates upon the whole. Rectangular grids have considerable utility since they serve to tie the linear dimensions of local and military locating systems as well as property boundaries into the highly controlled geodetic surveys.

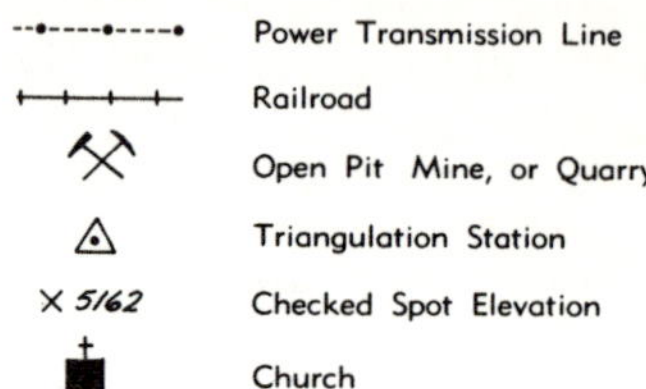

FIGURE 4.2.   Some Standard Map Symbols

the public land survey system

The Public Land Survey System is a rectangular one which was instituted by the Federal Government after 1785 to divide the public lands. Although it has never been completed and is therefore discontinuous, land units within it are outlined by red lines on many of the large scale topographic sheets issued by the U.S.G.S. Its principal use is for property description.

Public land surveys in a given area are referred to two primary lines, a *principal meridian* and a *base line* (see Diagram A, Figure 4.3). The first is a true meridian and the second a parallel of latitude. These lines are axes which intersect at the *origin* (Initial Point) of a system. The principal meridian is identified by a name or number which is used to reference any subdivision within its system. There are 34 systems, each with its own principal meridian in the United States, including Alaska.

Within a system, auxiliary base lines called *standard parallels* (correction lines) are spaced north-south along the principal meridian at 24-mile intervals. They are numbered by reference to the base line as, for example, the third standard parallel north, or the first standard parallel south. Similarly, *guide meridians* are spaced at 24-mile intervals east and west from the principal meridian along the parallels. These lines are referenced to the principal meridian by number as guide meridians east and west.

The spacing of the meridians and the parallels divides the land into tracts exactly 24 miles north-south and approximately 24 miles east-west. Because the meridians converge northward, the northern side of each

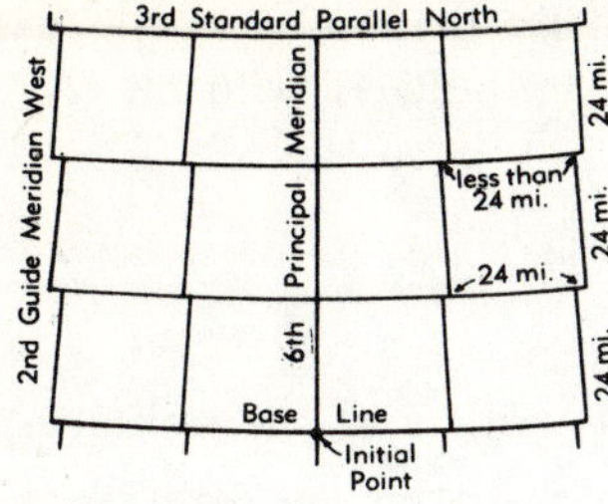

A. Division of land into 24 mile square tracts.

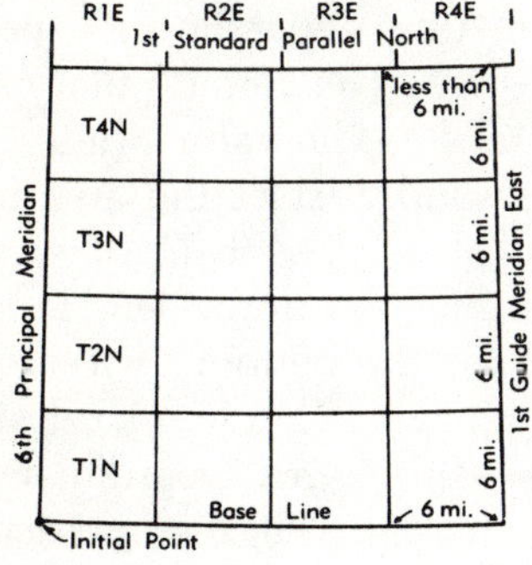

B. Division of tracts into 6 mile square townships.

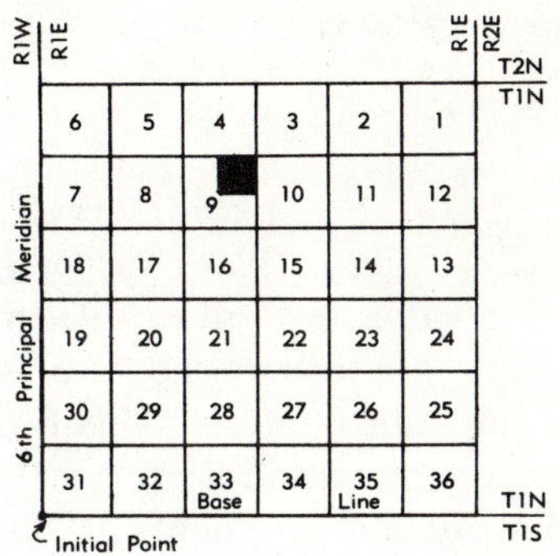

C. Division of Townships into 1 mile square sections.

FIGURE 4.3.   The Public Land Survey System

tract is slightly less than 24 miles long, although on the south the 24-mile dimension is held true.

To continue the division, each tract is further divided into 16 nearly rectangular land units by north-south trending *range lines* and east-west trending *township lines* (see Diagram B, Figure 4.3). These land units are called *townships*. Since the lines are spaced 6 miles apart, each township is approximately a 6-mile square. Townships are identified by reference to the lines as Township 2 (or some other number) North or South, Range 3 (or some other number) East or West. Each township is further divided into 36 squares, or *sections*, 1 mile on a side and containing 640 acres (Diagram C). Sections are identified by a numbering system beginning as Section 1 in the northeast corner, and ending with Section 36 in the southeast corner of the township.

The system permits accurate positioning of a property with a known area. Thus, the black square in Diagram C of Figure 4.3 is identified as the "N.E. 1/4 of Section 9, Township 1 N, Range 1 E of the 6th Principal Meridian." This is an accurate description of the location of a 160-acre land parcel, one-half mile square.

Figure 4.1 illustrates the application of the Public Land Survey System. It shows a portion of a Principal Meridian and parts of the section, range and township enumerations. To demonstrate the use of the grid in locating a point, a bench mark symbolized with an "X" and identified as BM 5192, is located in the S.E. 1/4 of Section 21, T4S, R1W.

other grid systems

Other rectangular grid systems, the state grids and the military grids also appear on the U.S.G.S. and the U.S.C. & G.S. topographic maps. Essentially, they are networks of squares, with their ordinates shown on the maps as blue or black tick marks which intersect the map margins. Ordinate values are given in feet or meters, depending on the system. Each square is referenced to a *false origin* (an initial point) from which east and north measurements are taken. The false origin, with coordinates Zero East and Zero North, lies southwest of the area of interest so that all measurements taken from it are positive. All *eastings* (easterly measurements) from the origin increase from left to right across the map, while *northings* increase from the bottom to the top.

State grids are made up of 1,000 foot squares. They are so designed that any point on them which is given a pair of grid coordinates can be located in reference to a triangulation station. They are particularly useful to the land surveyor using plane survey methods, for, although he works within the conception of a flat grid, his results must be related to the geographic grid as it is determined by geodetic methods. Each state has its own grid system which can be tied into the larger geodetic surveys. For each state data from the surveys are recorded on either a Lambert Conformal Conic or a Transverse Mercator Projection in its secant form, depending on the shape and orientation of the state. For example, Illinois has long N-S and shorter E-W dimensions and uses the Mercator. Ohio, which is compact in shape, uses Lambert's. In further contrast, Florida, which in its northern parts has a long E-W dimension and in the south has a long N-S extent, combines the two to produce the base for its rectangular grid. In all cases state grid coordinates are given in survey feet east and north from a false origin.

The Universal Transverse Mercator (UTM) Grid is designed for use on maps developed around the Transverse Mercator Projection in its secant form. However, it also appears on the common topographic map sheets of the U.S.G.S. and the U.S.C. & G.S. which are based on conic projections. The grid provides a map referencing system for all points on the earth between 84°N and 80°S. As in the state grids, any point defined in terms of the UTM Grid coordinates can be located further in relation to the geographic grid of latitude and longitude. The grid has two principal uses. First, because it provides continuous coverage for a large part of the earth's surface, it has general military utility. Second, since it is made up of a system of relatively small squares (1,000 meters), it is becoming more and more useful as a basis for statistical mapping in small areas such as cities and counties which are mapped at large scales.

The area covered by the UTM Grid is divided into 60 N-S trending *grid zones*, with each extending through 6° of longitude and 164° of latitude. Each zone is mapped on its own central meridian with two lines of exact scale, and is caused to overlap its neighboring zones. The lines of exact scale are spaced 180,000 meters east and west of the central meridian and are parallel to it. All grid zones are divided into 20 quadrilaterals, with the northernmost being 12° of latitude in length and each of the others 8° long. Each of the quadrilaterals is further divided into 100,000, 10,000 and 1,000 meter squares.

All of the quadrilaterals and their included squares within a zone are referenced to one of two false origins. Each of the origins is located 500,000 meters west of the central meridian of its zone. Consequently, the central meridian has an easting ordinate of 500,000 while the lines of exact scale have similar ordinates of 320,000 and 680,000. For Northern Hemisphere portions of a zone, the origin is located on the Equator which is given a zero meters north ordinate. For Southern Hemisphere portions, the origin is located 10,000,000 meters south of the Equator. This means that for all parts of a grid zone all eastings are positive and increase from left to right (west to east) across the map. Northings, too, are positive and increase from the bottom to the top (south to north) of the map. All values are stated in meters east and north of the origins.

The 1000 meter squares are shown on topographic maps of the U.S.C. & G.S. and the U.S.G.S., which are drawn to scales of 1:100.000 and larger. Figure 4.1 serves as an illustration. A square is identified on the map by giving the coordinates of the lower left-hand (southwest) corner. The co-ordinates of the corner are stated with eastings given first and northings given second in accordance with the rule: *read right up.* For example, Bench mark 5192 is located on the UTM Grid square 43 east, 35 north, about 600 meters east and 100 meters north of the lower left-hand corner. Standard practice designates the square in a kind of shorthand of 4335. The coordinates of the bench mark are then 436351. The third digit, the number 6, refers to the 600 meter easting while the final digit refers to the 100 meter northing.

When the UTM Grid is superimposed upon a conic projection, its lines diverge from the parallels and meridians on the map. This is, of course, the result of the fact that conic projections show meridians as converging lines while the north-south trending lines on the grid are parallel. The divergence is known as the *grid declination*. It increases with distance from the Equator, and is measured as the angle between *grid north* and a meridian. Grid north is the north indicated by the vertical lines of the grid. Where the UTM Grid is used, grid declination is commonly shown as a part of the declination diagram. The diagram is illustrated in Figure 4.1.

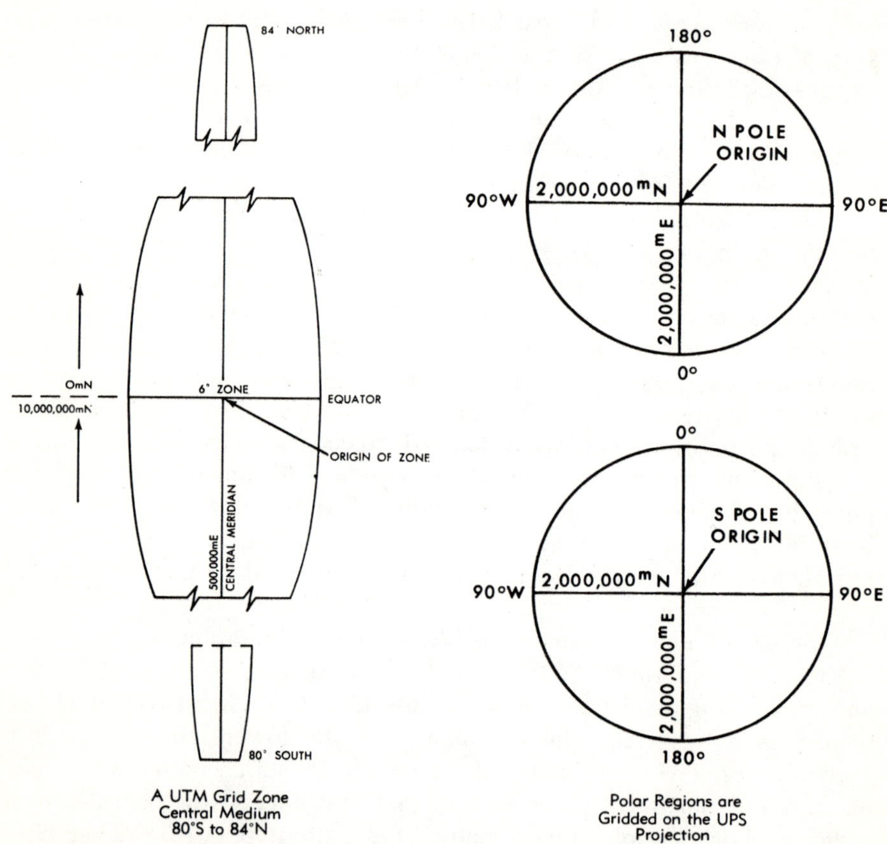

FIGURE 4.4. Elements of the Universal Grid System. (Courtesy of the Department of the Army, Field Manual, FM 21-26, **Map Reading,** Washington, D. C., 1965.)

Part II

The Earth
in the Solar System

Consideration of the earth as a member of the solar system is of fundamental interest in a study of general physical geography. Natural conditions everywhere on the global surface reflect the sun's influences and the earth's rotation and revolution. The annual and daily distributions of solar energy; the fact of day and night; and even the fundamental, more or less stable form of the earth: all of these depend in one manner or another upon basic earth movements and earth-sun relationships.

With such a point of view, the earth is here thought of as a part of a larger whole; a unit within a systematically organized grouping of interdependent and interacting bodies. It moves in response to the general laws of planetary motion in a clearly defined path in relation to the sun, and it rotates on its axis. It is to these movements that we relate many of the processes and factors which give character to various parts of the earth and contribute to the differentiation of its surface.

Planetary Motions

The earth is one of nine known planets in our solar system. All move through elliptical paths of revolution, or orbits. Within the orbits the sun is positioned at a common focus. Each planet rotates on its axis as it revolves around the sun. Although the planets vary considerably in size, their shapes are similar, being generally ellipsoidal with polar flattening. Their flattening varies from a maximum of about 1/9.5 (Saturn) to that of the earth which we have seen to be about 1/300.

Table 5.1 shows the earth to be the third farthest from the sun and the sixth largest of all the planets.

TABLE 5.1

Solar System Data

	Approximate Mean Diameter, Miles	Approximate Mean Distance from the Sun (1, 000, 000 miles)
Mercury	3, 100	36. 0
Venus	7, 700	67. 2
Earth	7, 918	92. 9
Mars	4, 215	141. 6
Jupiter	86, 800	483. 4
Saturn	71, 500	886. 2
Uranus	31, 700	1, 782. 0
Neptune	31, 000	2, 792. 4
Pluto	8, 000 (?)	3, 671. 2
Sun	864, 000	

Source: Mehlin, T. G. *Astronomy*, p. 385, John Wiley and Sons. 1961. (By permission)

Planetary Revolution

The fact that planets move was a phenomenon well-known to the ancients. Observers of antiquity saw that while many thousands of stars held their same relative positions in the heavens for long periods of time there were other objects, as well as the sun and the moon, which appeared to wander in peculiar backward and forward (retrograde) paths among the stars. These they labeled planets, a term meaning *wanderers*. The true nature of their paths, however, was not understood until early in the seventeenth century. In 1620 their orbits were described by the great astronomer Johannes Kepler when he published his now famous *Laws of Planetary Motion*.

Kepler's Laws were strictly empirical. They were derived through analysis of a lengthy series of observations made previously by Kepler's teacher, Tycho Brahe, and did no more than describe certain characteristics of the movements of planets around the sun. Their explanation had to wait the development of Sir Isaac Newton's *Laws of Motion and Gravitation*. These laws, published late in the seventeenth century, are used today to explain the shapes of planetary orbits and the variable velocities of the planets as they move around their primary, the sun.

Kepler's laws

Kepler's observations established three basic facts regarding the orbits of planets. Of these, two have particular significance to geographical study. When applied to the earth they describe several features of earth-sun relationships which affect the distribution of solar energy over the earth's surface, and our measures of time. These two laws are discussed below as the *Elliptic Law* and the *Law of Equal Areas*.

The Elliptic Law states that a planetary orbit is an ellipse, with the sun located at one focus. This means that as a planet moves through its orbit its distance from the sun varies. When at the point nearest the sun the planet is said to be in *perihelion* (from the Greek, *peri*, meaning near; and *helios*, meaning sun). When farthest from the sun it is said to be in *aphelion* (from the Greek, *apo*, meaning away from). The mean distance between a planet and the sun is one-half the sum of the perihelion and aphelion distances. This is the semi-major axis of the ellipse. The shape of an orbit is defined by reference to its eccentricity which is measured by the difference between perihelion and aphelion distances divided by their sum. Since the eccentricity of a circle is zero, low value eccentricities indicate nearly circular orbits.

The Law of Equal Areas states that a line drawn from a planet's center to the center of the sun sweeps across equal areas within the orbit in equal intervals of time. This means that the velocity of a planet in its orbit varies (see Figure 5.1). When at perihelion the planet moves at a greater velocity than when at aphelion. In general it may be said that the greater the eccentricity of an orbit, the greater the variation there is in the orbital velocity of the planet. Thus, in the case of Pluto which has an orbital eccentricity of 0.250, the orbital velocity of the planet varies from 3.8 to 2.3 miles per second. In contrast, the earth with an orbital eccentricity of about 0.016, has an aphelion velocity of 18.2 miles per second while its velocity at perihelion is 18.8 miles per second.

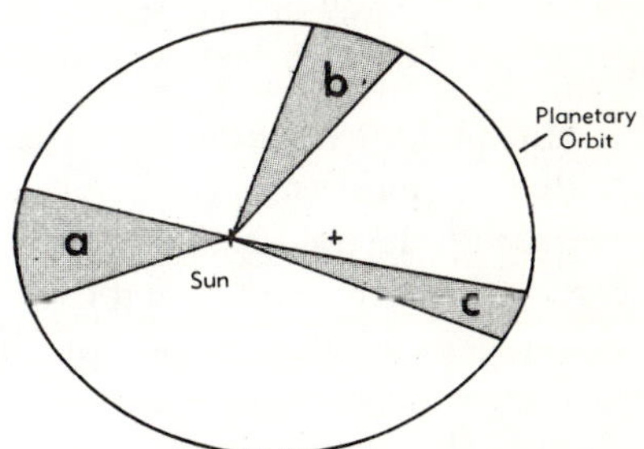

FIGURE 5.1. Kepler's Law of Equal Areas. A line drawn from a planet's center to the center of the sun sweeps through equal areas in equal lengths of time. Areas a, b, and c are equal. The arc segments bounding their sides are unequal. Thus the planet moves through arc segments of unequal lengths in equal periods of time and its velocity in orbit varies.

Newton's laws

Newton's explanations of elliptical orbits relied upon an analysis of the combined effects of two basic factors: gravity, and the tendency of a body in motion to continue in the same straight line. Kepler and others had discussed gravity by noting that the celestial bodies have the power of attracting. Galileo had refined the idea somewhat by demonstrating that all bodies fall, (are attracted to the earth) at the same rate, regardless of their size. Moreover, he had suggested the now familiar principle that a body in motion will continue to move in the same straight line unless acted upon by an outside force. Newton was able to crystallize these and other ideas into several basic principles, or laws, and so explain Kepler's observations.

In the application of Newton's principles as they affect planetary motions, we need to remember that an object moving through space is not affected by the friction which would result if an atmosphere were present. This means that once put in motion, such an object will continue to move along a straight line through the heavens until it is acted upon by another outside force. If the other outside force is continuously applied at an angle to the straight line path the resulting direction of movement will be a curve. The object will, however, have a tendency to resume its straight line path. It will tend to move outward along a line tangent to the curve. The measurement of this tendency increases as the radius of the curve is shortened.

In the case of the sun and a planet, or a planet and its moon, the outside force is the gravitational attraction between the two bodies in question. The magnitude of the gravitational attraction is expressed as follows:

$$F = GMm/d^2$$

Where: M and m are the masses of the bodies in question; G is the gravity constant, which is used to provide numerical equality between the various measures used; d is the distance between the two bodies; and F is the force of gravity.

In qualitative terms this means that if the *mass* of either of the two bodies is increased the gravitational attraction between them is also increased. If the distance between them is increased the attraction is lessened. Furthermore, it bears emphasis at this time that regardless of differences in the masses of the two bodies involved, each exerts an attraction on the other. Bodies of smaller mass tend to move more readily toward the larger.

It is important that the concept of mass be correctly understood. Mass is, in the most simple of terms, a measure of the amount of matter in a body exhibited at a point which is the center of gravity, or the center of mass. The gravitational attraction between any two spherical and homogeneous bodies is exerted along a line connecting their mathematical centers.

A planet in orbit may first be thought of as a spherical, homogeneous body which moves in a straight line path through the sky. It is influenced only by the sun, also spherical and homogeneous. At some point the planet is attracted toward the sun. Since the planet is much less massive than the sun it will tend to "fall" toward that body. In falling it will follow a curving path as the result of its tendency to continue in a straight line on the one hand and its tendency to fall toward the sun on the other. As it falls its velocity increases because its distance from the sun is decreasing. In consequence, both the tangential force and the gravitational attraction increase until, at some point, they are equal. Since there is no atmosphere in space to oppose this motion, the planet will continue to follow a curved path which closes on itself around the sun.

If we now return to our original assumption, that is, an object moving along a straight line, it is not difficult to imagine two extreme sets of conditions. The one in which attraction is not strong enough to draw the object to the sun would result in no more than a deflection of the object.

Its straight line path would only be "bent" and the curve would not close on itself. The other would cause the object to fall directly into the sun. Between the extremes there are two other possibilities. The planet will settle into either a true circular path or an elliptical one. The full explanation of the conditions which produce the one or the other is beyond the scope of this book. We need only say here that, for all of the planets in our solar system, the orbits are ellipses, along which the sun's satellites move with varying velocities.

Planetary Rotation

That the earth rotates on its axis is a commonplace that is demonstrated every day by the apparent east-to-west motions of the sun and the stars through the sky. Moreover, a similar rotation of the larger and closer planets can also be observed. Telescopic observations clearly indicate the rotations of Mars, Jupiter, and Saturn as their visible surface markings shift from one side of the planetary disc to the other and finally disappear on the far side. For others, special observational techniques are used to determine rotational speeds. Rotation, then, is a common feature of planets. It can be demonstrated for their moon satellites as well.

Rotation has two direct effects upon a planet. In the first place it produces the equatorial bulging and related polar flattening that are common to all planets. As we have seen in the case of the earth, this affects distance measurements along the lines of longitude. Secondly, rotation creates a tendency for the planetary axis to retain its attitude, or orientation, in space. In a sense it prevents an orbiting body from rolling about in a random fashion as it moves in its revolution.

A planet in rotation may be thought of as a top. If the top is set to spinning, and a slight force is applied laterally to the extension of its axis, it will resist movement. If the axis is tilted slightly, and the tilting force removed, the top will oscillate slightly and settle into a new position. Further, if the spinning top is thrown into the air its axis will continue to point in the same direction. This gyroscopic effect tends to maintain the direction of the top axis in relation to some distant point. In the case of planets, observations show that their axes of rotation tend to retain constant orientations for long periods of time. The inclinations of the axes to the planes of the orbits remain almost unvarying. Some variations, however, have been noted. They relate in part to the generally oblate form displayed by all planets.

As a planet rotates the individual particles of which it is composed tend to move outward from the rotational axis. For individual particles these tendencies partially overcome their gravitational attractions for the gravity center of the body. The net result of the interplay of forces is that the diameters of what may be thought of as latitude circles are stretched. Maximum stretching takes place along the equatorial diameter, midway between the planetary poles. At the poles themselves, gravity alone is operative. Particles located at the poles are attracted only toward the center of gravity along the axis. The result is that the axis is shortened and an equatorial bulge, with polar flattening, develops. All planets and their moons as well are characteristically oblate, or ellipsoidal, in form.

The development of the ellipsoidal form of a planet may be further described mathematically. For our purposes it may be better visualized as follows. If, for example, a series of wells were drilled all along a meridian circle of a planet, and each well extended to the planet's gravity center, and connected at that point, the water level in each would be the same. This generalization would hold for a stationary and precisely spherical planet. If, however, the planet were then set to rotating, the water would rise in those wells located on the equatorial diameter. For wells at the poles the level would drop. For other wells positioned at in-between points the water surface would assume in-between levels. A line drawn to connect the various water levels along a meridian would describe the outline of the planetary ellipsoid.

The distortion of the planetary sphere and the development of the ellipsoid has the effect of redistributing the mass of the sphere. As a result, the attraction of the sun or a planet's moon for various parts of the planet vary, as shown in Figure 5.2. The figure shows an hypothetical planet and two positions of its moon. In position number 1, the planetary axis is inclined to a line connecting their centers of gravity. This connecting line represents a "side view" of the moon's orbit around the planet. At *A* the moon's attraction for the planet is greater than it is at *B*, simply because *A* is closer to the moon than *B*. Because of the difference there is a tendency for the planetary axis to be turned into a position at right angles to the orbit. This tendency is resisted, but not completely overcome, by the gyroscopic effect described previously. Quite clearly no such tendency would exist if the planet were a perfect sphere, or if the moon were in position number 2, directly above the equatorial bulge.

A moon revolves around its planet as the planet revolves around its sun. In general, for any moon-planet pair, the orbital velocities of the two are different and their planes of revolution are inclined to one another. Consequently, the relative positions of the three bodies are constantly changing. The results are that the magnitudes and directions of forces which act to reposition the rotational axis also change, and the axis describes a very complex motion. The axis oscillates much as a top does when subjected to varying lateral forces. Primarily, as shown in Figure 5.2, it describes a cone with apex at the planet's center, over a long period of time. This oscillation is known as the precession of the axis. Secondarily, over a much smaller period of time, the axis oscillates very slightly in relation to its average position. The most important of the secondary oscillations is known as the *principal nutation*.

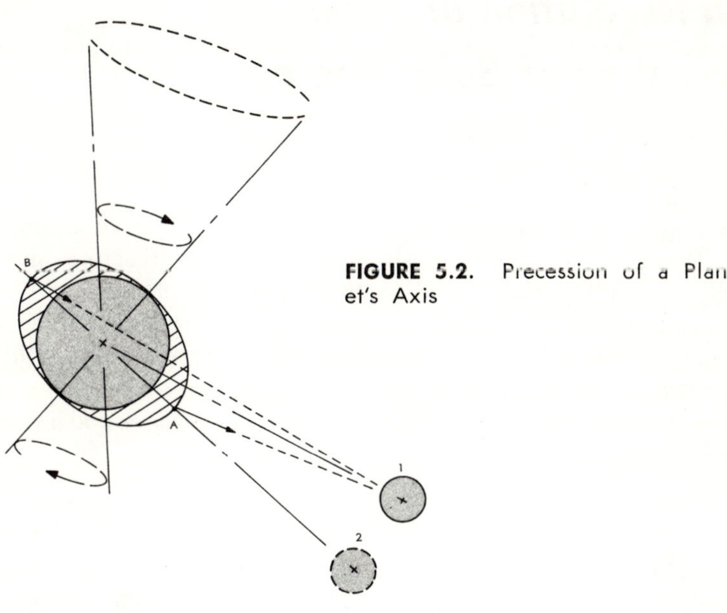

FIGURE 5.2. Precession of a Planet's Axis

Earth Revolution and the
Distribution of Solar Energy

The revolution of the earth around the sun has several important effects. Primarily, it influences total solar energy receipts and their distribution over the globe; and in this sense it contributes directly to the differentiation of the earth's surface. Also, because it serves as the great celestial calendar while providing lines of reference within our basic locational system, it functions as a framework of sorts for much of human activity. It thus becomes a factor of considerable interest to geographical study.

The Earth's Orbit

The earth's orbit, like that of other planets, is an ellipse, with the sun located at one of its foci. Its major axis is approximately 186 million miles long and the distance between its foci is 3 million miles. These measurements, which yield a calculated eccentricity of 0.016, indicate that it is very nearly circular. By comparison and as a matter of general interest, the eccentricities of other planetary orbits range from the smallest, 0.007 for that of Venus, to 0.250, the largest, in the case of Pluto.

The eccentricity of the earth's orbit expresses the degree to which the sun is offset from the center of the ellipse. Because the sun is offset its distance from the earth varies as that body revolves around it. Since perihelion distance is 91.5 million miles, and aphelion distance 94.5 million miles, the variation amounts to 3 million miles, or about 3 per cent. The mean earth-sun distance is 93 million miles.

The circumference of the orbit is about 583 million miles; the time interval, or period, required for the earth to make one revolution through it is one year. Consequently, the earth's orbital velocity averages an incredible 66,600 miles per hour, or about 18.5 miles per second. Although it is probably unnecessary to dwell on the fact that the period is one year, the meaning of the term "year" needs some clarification at this point. It will be dealt with in more detail in later sections of this book which consider the general problem of time and its measurement. The year referred to is the so-called tropical year, the time interval between two successive spring equinoxes. In terms of days it averages 365 1/4 mean solar days of 24 hours each.

some effects of ellipticity of the orbit

The elliptical form of the earth's orbit has two principal effects. In the first place, as we have seen, it affects orbital velocity. It causes the earth to move at different rates of speed in different parts of the orbit at different times of the year. In the second place, it influences the amount of solar energy or insolation received on the earth at different times of the year. This results from variations in the earth-sun distances which relate to the offset position of the sun.

The earth's velocity in orbit referred to above is an average figure. However, as we have seen from Chapter 5, it varies between 18.2 and 18.8 miles per second. The variation is sufficient to influence any precise statement of time which is based on the earth's period of revolution. This influence is discussed in the chapter on time and its measurement.

The earth reaches its perihelion on January second or third, and aphelion on the fourth or fifth day of July. This means that the earth and the sun are closest together in the Northern Hemisphere winter, and farthest apart during its summer. Since receipts of solar energy amount to as much as seven per cent more at perihelion than at aphelion, this means further that during the northern winter more of the sun's energy is being received on the earth than at any other season.

The extent to which seasonal differences in total insolation receipts relate to general patterns of earth surface temperatures has never been fully determined, simply because temperature on the earth depends upon so many other variables. Chief among the variables are the uneven distribution of land and water over the earth, the albedo, or reflective power of various surfaces, and atmospheric conditions. Theoretically, some relationships between variations in solar energy receipts and earth temperature should exist, and they might very well be studied in small areas. Our primary concern here, though, is with the world-wide patterns of insolation to which some later study may refer. We are less interested in total energy receipts on the earth than we are in obtaining answers to the question of just how it is that the energy available is distributed over its surface.

General Features of Solar Energy Distribution

Three factors play primary roles as determinants of the distribution of solar energy over the earth's surface. Rotation and revolution in the presence of a relatively motionless sun distribute the sun's rays and its energy in orderly patterns on the globe. Details of the patterns are determined by the angle of inclination of the earth's axis to the plane of the orbit.

At the outset it is necessary that basic locational relationships between the earth and the sun be correctly visualized and that certain assumptions are understood. Initially, we assume that the earth is a sphere and that all the sun's rays which strike it are parallel. For justification of the spherical assumption the reader is referred to earlier sections of this book which deal with the earth's figure. Supposition of the parallelism of the sun's rays is justified by consideration of the extremely great distance between the earth and the sun as it compares with the earth's diameter. If, for example, two sun's rays originating at the same point on the sun are projected to the earth, with one made tangent near the North Pole and the other made tangent near the South Pole, the angle between the two would be so small as to be hardly measurable. In numerical terms it would be an angle with a tangent equal to 8000 (the earth's diameter) divided by 93,000,000 (the mean earth-sun distance). Any error made by assuming this angle to be equal to zero (as is the case when the two rays are considered to be parallel) is negligible in view of the initial assumption that the earth is a smooth-surfaced sphere.

In our examination of locational relationships we first recall that the earth rotates on its axis while revolving in orbit around the sun. Secondarily, we note that the axis of the earth intersects the plane of the earth's orbit and that the directions of rotation and revolution are the same when the earth, its orbit, and the sun are viewed together. If we imagine ourselves to be at some point from which we can look "down" on the North Pole, the earth will be seen to rotate on its axis and revolve in its orbit in the same counter-clockwise direction. Since the earth can now be seen to rotate "under" the sun, it is apparent that rotation has the effect of distributing sunshine longitudinally over its surface (Diagram A, Figure 6.1).

A third set of relationships has to do with variations in the energy received at various places on the lighted part of the earth. The sun, with a diameter more than 100 times that of the earth, bathes one-half of the globe in sunshine, leaving the other half dark, much as one side of a small ball would be lighted if it were held stationary in the presence of a point source of light (Diagrams B and C). The two halves of the globe are separated by the *circle of illumination;* the dividing line be-

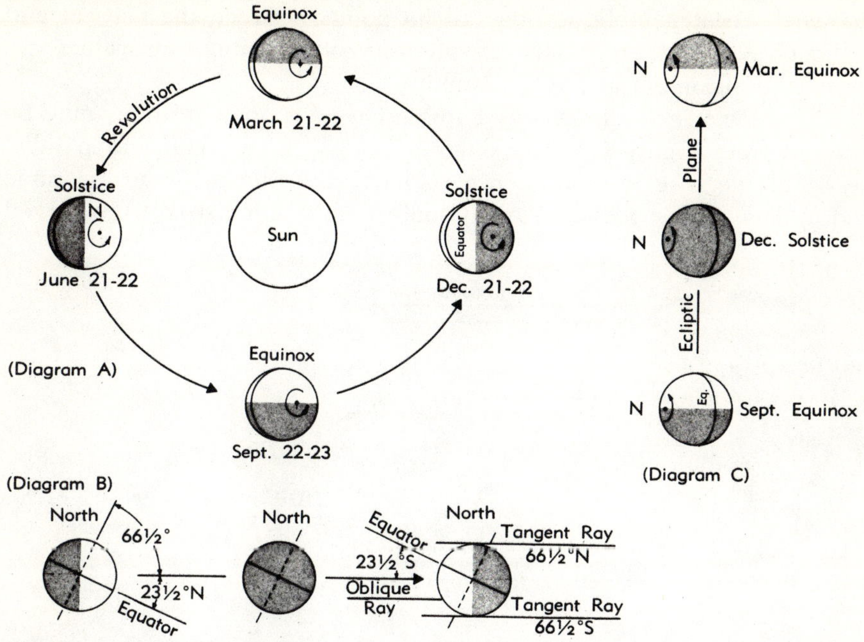

FIGURE 6.1. Earth-Sun Relationships

tween the day and night sides of the earth. The plane of the circle intersects the plane of the earth's orbit at a right angle. As previously noted the circle of illumination is a great circle which bisects the equator (Chapter 2).

Because the earth's surface is curved, while all sun rays are straight and parallel, the insolation received on the daylight half of the earth varies from place to place. Individual sun rays falling on the curved surface of the hemisphere strike the surface at different angles. The intensity, or concentration of solar energy per unit of area, varies accordingly. If we neglect for the time being the refracting (bending) effect of the atmosphere on a sun ray, its intensity is determined locally by the angle it makes with the curved earth's surface. Intensity is greatest where the angle is 90°, with the ray vertical. It is least where the angle is zero, with the ray tangent. It has an in-between value where the ray is oblique. We can, then, point to zones of difference in the intensity of insolation at a given time, and discuss the various related sun's rays as tangent rays, oblique rays, and the vertical ray. Tangent rays mark points on the circle of illumination where energy receipts are minimal. The one vertical ray strikes the earth directly below the sun, at the *subpoint* of that body where the intensity is maximized. If extended it would pierce the earth's center. Oblique rays fall on points between the vertical and tangent rays. At these points the intensity of insolation has values between the maximum and the minimum.

Differences in the intensity of insolation at various locations may be inferred from Figure 6.2. The diagram shows three equal and parallel rays. For the sake of convenience each ray is assumed to be the same and to have a cross-sectional area equal to unity. Each strikes the curved

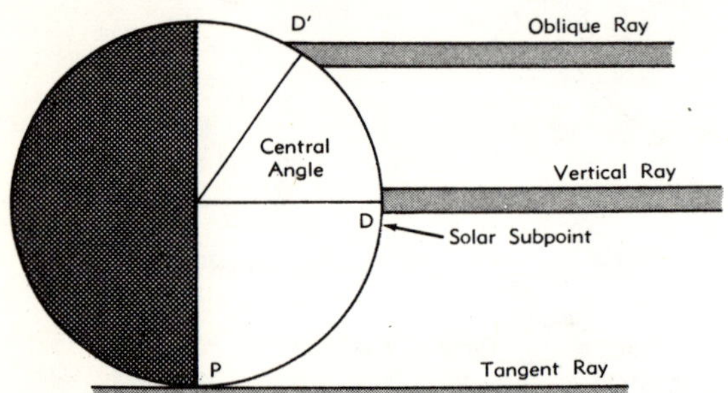

FIGURE 6.2. Intensity of the Sun's Rays

earth's surface at a different place and at a different angle. The energy of the vertical ray is concentrated in an area represented by the distance D. An oblique ray delivers the same amount of energy to an area represented by D'; a tangent ray touches the earth at point P. Inspection indicates that D' is greater than D, and that P is a point with no dimensions. Consequently, the intensity, or concentration of solar energy per unit of area, decreases away from the vertical ray. In mathematical terms, the intensity decreases as the cosine of the central angle between the vertical ray and other rays decreases.

At this point it must be recalled that there is but *one* vertical ray. The earth curves away from it in all directions, and in consequence, oblique and tangent rays fall on the surface all around it and in all directions from it. Thus, the central angle referred to above may be measured in the plane of any great circle which passes through the sun's subpoint.

A fourth set of relationships is established when we change our position and observe the earth-sun complex from a point directly "above" the Equator (Diagram B, Figure 6.1). From such a viewpoint the elliptical earth orbit appears as a straight line and we are given a "side view," so to speak, of the plane of revolution, or the *plane of the ecliptic*. This line may also be thought of as a representation of the vertical ray of the sun. The vertical ray is always coincident with the plane and, if extended, will pass through the centers of both the sun and the earth. The latitude upon which it falls is the *declination* of the sun.

The side view also shows that the earth's axis is inclined at an angle of 66°30′ to the plane of the ecliptic. Because this angle remains fixed throughout the year, the axis is at any time parallel to all positions it has assumed or will assume during the period of revolution. As a result, the vertical ray, which is coincident with the ecliptic plane, moves slantingly across the earth's surface during the earth's passage through its orbit. Stated differently, although the sun remains in place during the year, its declination changes as the earth moves around it, as the direct result of the inclination and the parallelism of the earth's axis. The sketch shows that the vertical ray shifts from a northerly to a southerly latitude in about one-half a year.

Because the angle of inclination is 66°30′, the highest latitude ever reached by the ecliptic plane and the vertical ray is 23°30′ (90° minus 66°30′). This is the latitude marked by a special parallel, a *tropic*. In the Northern Hemisphere it is the Tropic of Cancer, and in the Southern it is the Tropic of Capricorn. The tropics are defined as the highest parallels of latitude which receive the intense insolation associated with the vertical ray.

seasonal changes in solar energy distribution

At this point it becomes apparent that during one annual revolution of the earth around the sun the vertical ray shifts from one tropic to the other and back again. In doing so it will cross the Equator twice. During this migration the circle of illumination, which is always perpendicular to the vertical ray, will also shift its position in direct relation to changes in the sun's declination. It is to these shifts that we attribute the redistribution of solar energy which is associated with change in the seasons. In general terms, the earth revolves through its orbit in one year. Four times during the year the vertical ray is located at critical latitudes, the latitudes of the Equator and the tropics. It is these times or days that mark the beginnings and ends of the seasons.

At two times during the year the vertical ray falls on the Equator. At these times, March 21-22 and September 22-23, the circle of illumination passes through the poles and cuts all parallels in half (Figure 6.3). This means that at all latitudes one-half of the surface is experiencing day and one-half is experiencing night. Days and nights are equal in length all over the earth. These are the times of the so-called *equinoxes*.

About three months (one-fourth of a year) after the Northern Hemisphere spring (vernal) equinox, the vertical ray reaches the Tropic of Cancer at 23°30′ N. At this time, June 21-22 (the Northern Hemisphere *summer solstice*) the circle of illumination, too, has shifted its position by 23°30′ and in such a way that some of the sun's rays pass over the

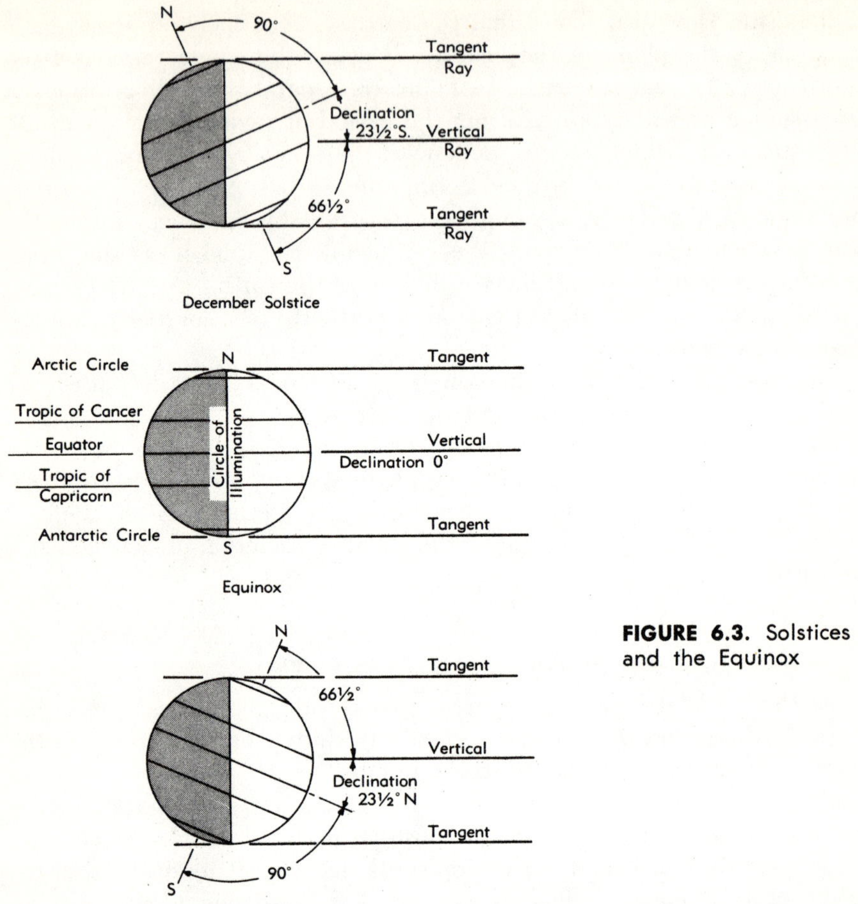

FIGURE 6.3. Solstices and the Equinox

North Pole and are tangent at latitude 66°30′ N. The North Pole now lies between a point on the circle of illumination and the vertical ray. It is, in effect, tilted toward the sun and all latitudes greater than 66°30′ N are exposed to the sun for 24 hours, or one rotational period. In contrast, in the Southern Hemisphere, the pole is tilted away from the sun, and all latitudes greater than 66°30′ S experience a 24-hour night. Like the tropics, the 66°30′ parallels are special parallels, the so-called *circles,* known as the Antarctic Circle in the Southern Hemisphere and the Arctic Circle in the Northern Hemisphere. They mark the lowest latitudes that ever receive the minimal insolation associated with the sun's tangent rays.

After the June solstice the vertical ray begins its migration southward, arriving at the Equator on the day of the northern fall equinox (vernal equinox for the south). Continuing southward it reaches the Tropic of Capricorn on December 21-22, the southern summer solstice. At this time the situation described above for the Northern Hemisphere is reversed, with all southerly latitudes greater than 66°30' S being exposed to 24 hours of sunshine, and all comparable northerly latitudes experiencing darkness for the same period.

From the foregoing it is evident that there is an orderly seasonal change in the lengths of day and night at any latitude. The declination of the sun and the position of the circle of illumination change from day to day and from season to season and in consequence the duration of sunlight at all latitudes also changes. The exception to this is at latitude zero, where days and nights are always equal,[1] simply because it is here that the circle of illumination bisects the great circle of the Equator.

Latitude Determination by Sun Observation

A latitude determination from sun observations can be carried out in two steps. At noon, with the sun directly above the observer's meridian, the sun's altitude is measured as the angle between the sun and the horizon. The altitude is then subtracted from 90° to determine the *difference* between the observer's latitude and the latitude of the vertical ray (the declination). This difference is then applied to the declination to obtain the observer's latitude. In applying the difference, three cases are possible.

1. If the sun's vertical ray falls between the equator and the observer, the difference is added to the declination.
2. If the observer is located between the equator and the vertical ray, the difference is subtracted from the declination.
3. If the equator lies between the observer and the vertical ray, the declination is subtracted from the difference.

Preliminary working approximations of the relative positions of the vertical ray, the equator and the observer are based on the direction which the observer faces to see the sun as well as the sun's altitude. The declination for each day of the year is listed in the *Nautical Almanac* and is shown on the *analemma*. (See App. A).

---

[1]Some exceptions to this statement as well as to other statements regarding the lengths of days of various latitudes are normal and vary from year to year.

The Earth's Rotation

The earth's rotation, like its revolution, produces several important effects. It is a fundamental cause of the earth's oblateness and the variable relationships between angular and linear measurements of latitude. The principles involved are discussed elsewhere. By no means of lesser importance are the relationships which rotation and its accompanying precession have to problems of determining location on the earth. This and the related problem of the Coriolis Effect are considered in the paragraphs below, following a preliminary consideration of rotational velocity.

The Velocity of Rotation

The velocity of rotation may be expressed in two ways; either in terms of the angle through which the earth turns in a period of time, or as the velocity of a point on the earth's surface. In the first case it is called *angular velocity,* a measure expressed in terms of degrees of turn per minute, hour, or the like. To obtain the angular velocity of the earth, we divide 360° by 24 hours, the period of one rotation, and derive an answer of 15° per hour. At any given time the angular velocity for all points on the earth's surface is the same.

Linear velocity is the velocity of a point as it follows a circular path around the globe. Its value varies with the developed length, or circumference, of the circle of latitude upon which the point is located. It may be expressed in feet or meters per second, or in miles or kilometers per hour. It is, of course, derived by dividing the circumference of the circle of latitude upon which the point is located by 24 hours to obtain a velocity per hour, or by a multiple of 24 to obtain a velocity in some other temporal terms. Since the circumferences dealt with vary with latitude, the linear velocities of points at different latitudes also vary.

The linear velocities of rotation of points located at various latitudes on the earth are listed in Table 7.1. They are based on the circumferences of the circles of latitude on a spherical earth with an equatorial circumference of 24,901 statute miles and a day 24 sidereal hours in length.

TABLE 7.1

*Linear Velocities of Rotation
at Various Latitudes*

Latitude degrees	Velocity miles per hour
0	1038
30	899
60	519
90	0

Precession and the Orientation of the Earth's Axis

We have so far assumed that the orientation of the earth's axis is constant. If, however, the fact of the precession is taken into account, this generalization can be shown to be no longer true, except for relatively short periods of time. Over long periods precession causes the orientation of the axis to shift with relation to the stars.

The effect of precession on the orientation of the axis is shown in Figure 7.1. The figure shows that the extension of the northern endpoint of the axis very nearly coincides with Polaris, the North Star. Polaris, then, has a declination of approximately 90°, and provides a northerly reference. Also, it will be noted that the declination of Polaris will change as the axis precesses. The precessional motion changes the orientation of the axis by causing it, over a long period, to describe a cone. Since for the earth the apex angle of the cone is 47°, the declination of Polaris will change by that amount during one-half of the precessional period. Moreover, because the yearly rate of precession is 50 seconds of arc on the precessional circle, the period is about 26,000 years.

As is the case with all planets, the gravitational forces which cause the precession of the earth's axis are not constantly applied. This follows from the fact that when either the sun or the moon is directly over the equatorial bulge its precessional effect is nullified. The nullification of the precessional effect is periodic and introduces into the entire movement predictable fluctuations, or nutations. In the case of the sun pre-

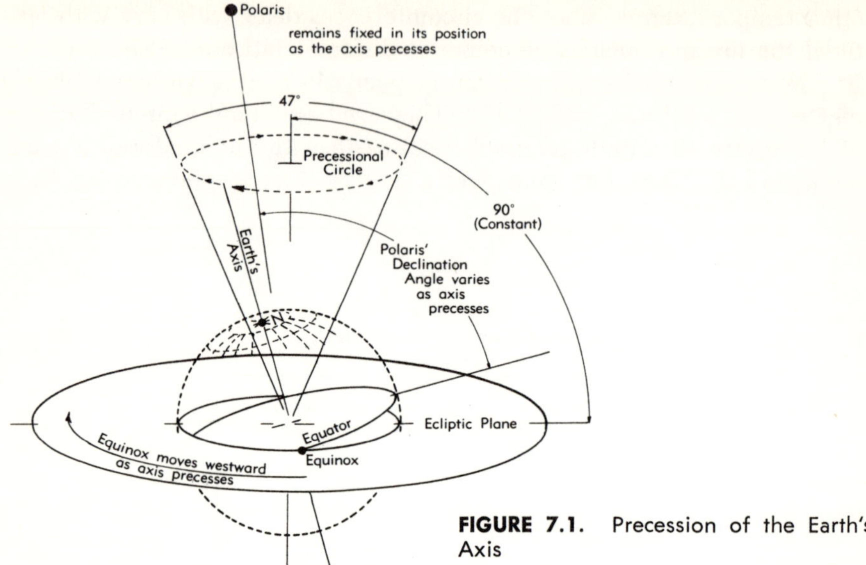

FIGURE 7.1. Precession of the Earth's Axis

cession ceases at the equinoxes, twice a year. At these times the sun's declination is 0°, a position comparable to position number 2 in Figure 5.2. For the moon this condition occurs 24 or 25 times a year because its orbit is inclined to the plane of the ecliptic. Inclination of its orbit causes the moon to cross the equator, thereby nullifying its precessional effect, two times during each of its monthly revolutions around the earth. (See Figure 8.4.)

Precession and nutation have no effect on the orientation of the ecliptic plane. The relationship of the ecliptic to Polaris remains the same. The Equator, however, moves with the axis. The more or less constant movement of the axis is then accompanied by a corresponding motion of the Equator which remains at right angles to it. As a result, the intersections of the ecliptic plane and the plane of the Equator (the equinoxes) move slightly as the axis precesses (Figure 7.1). Precession is such that the equinoxes move westward around the earth.

Latitude Determination and Polaris

Since Polaris remains approximately in place above the North Pole for relatively long periods of time, it provides a point of reference for directional and latitude determinations. When observed at night the entire panorama of the sky shifts westward as the result of the earth's eastward rotation. Above the rotational center at the Pole, however, there is no shift. Since its declination is nearly 90° N, Polaris describes a relatively small circle in the sky while the remaining elements of the firmament appear to move through larger circles. Thus the Polestar provides us with an approximate northerly reference. Our approximate latitude at any point in the Northern Hemisphere is equal to the star's altitude, or the angle between the horizon and the star (see Figure 7.2).

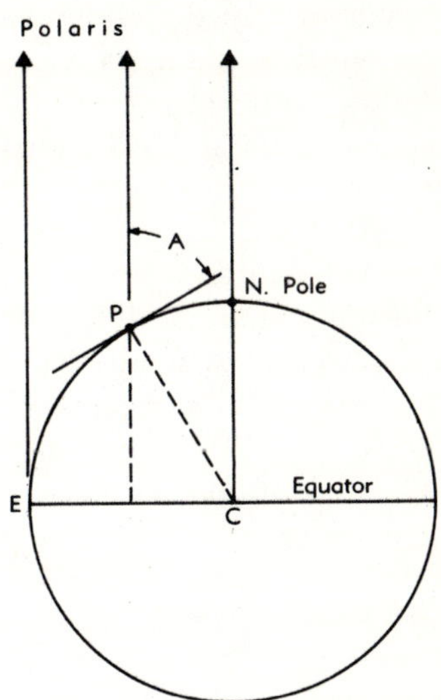

FIGURE 7.2. Determination of Latitude from a Polaris Observation

Angle "A", the altitude of the Polestar, equals angle PCE, the Latitude of point P.

Several factors combine to complicate the relatively simple determination of latitude as shown in the sketch. First, we recognize that Polaris is not directly above the Pole. At the present it is offset from the Pole by about 1°. This means that Polaris observations taken from the same place at different hours of the day will change slightly as the earth rotates. Therefore, any observation of the star's altitude which is made from the earth's surface must be corrected in such a way as to eliminate error made by assuming the ideal conditions. Corrections for the observed altitude of the star, as well as those made for the longitude of the point in question and the time of the observation, are given in *The American Ephemeris and Nautical Almanac*. The *Ephemeris* is published annually by the U.S. Naval Observatory in Washington, D.C.

The Coriolis Effect

Rotation also causes the so-called Coriolis Effect; the apparent deflection suffered by a body moving across the earth's surface. In the Northern

Hemisphere the deflection is to the right, or clockwise. In the Southern it is to the left, or counterclockwise. Deflection results from the movement of points and lines on the earth grid under the path of the moving body. It is illustrated by The Foucault Pendulum Experiment which is often used to prove the earth's rotation.

A heavy iron ball is suspended by a long wire in such a way that it can swing in any direction. It is caused to swing through a definitely oriented vertical plane. If it continues to swing the plane will change its orientation with relation to the earth grid at a regular hourly rate as the day progresses. This is attributed to two factors: 1) The orientation of the plane remains fixed in relation to a distant star, and 2) A portion of the earth's surface beneath the pendulum turns around its own center as the earth rotates.

The maintenance of the orientation of the plane of swing follows from the well-known physical law which states that a body in motion tends to remain in motion along a straight line unless acted upon by an outside force. With respect to the earth turning beneath the pendulum, further explanation follows.

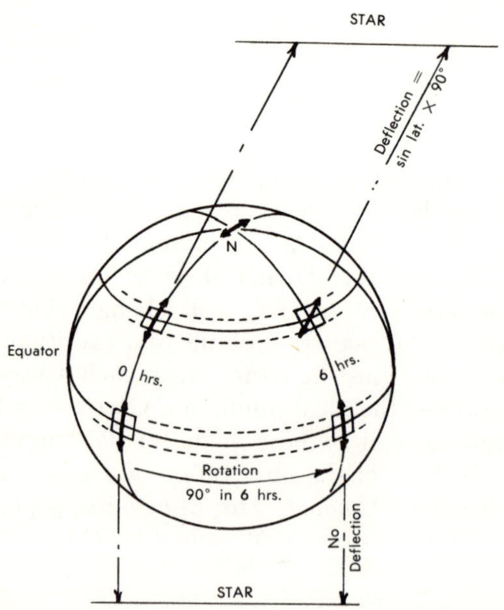

FIGURE 7.3. Deflection of The Foucault Pendulum

Let us consider three small squares at different latitudes. One is centered on a mid-latitude in the Northern Hemisphere. In 24 sidereal hours the entire square will move completely around the earth once. The southern side of the square however, will move farther than the northern side because it is farther from the earth's axis than is the northern side. The parallel along which it moves is slightly longer than the parallel along which the northern side moves, and its linear velocity is greater (Table 7.1). In consequence of this difference the small square will rotate around its own center. A square centered on 90° N will rotate around the pole. Since the earth rotates eastward, counterclockwise when viewed from above the North Pole, the small square in either case will also rotate *counterclockwise* around its center. This means that in the Northern Hemisphere the plane of a pendulum suspended above the center of the square will appear to change its orientation and rotate *clockwise* as the day progresses. A square centered on the Equator will not rotate and a pendulum suspended above it will not be deflected.

The rate at which the pendulum plane changes direction varies systematically with its latitude. If the pendulum is suspended above the North Pole its plane will change direction by 360° in one sidereal day. If suspended above a mid latitude the deflection will be somewhat less. In either case, the small square will turn counterclockwise and the plane of the pendulum will be deflected clockwise at the *same* rate. If located at the Equator, the square will not turn and the orientation of the plane will remain unchanged. The hourly rate of turn is equal to the sine of the latitude times 15°. (The sine increases from zero at lat. 0° to 1.000 at 90° and the hourly turn varies accordingly.) In six hours, the *total* deflection of the plane is sine latitude times 90°. For locations in the

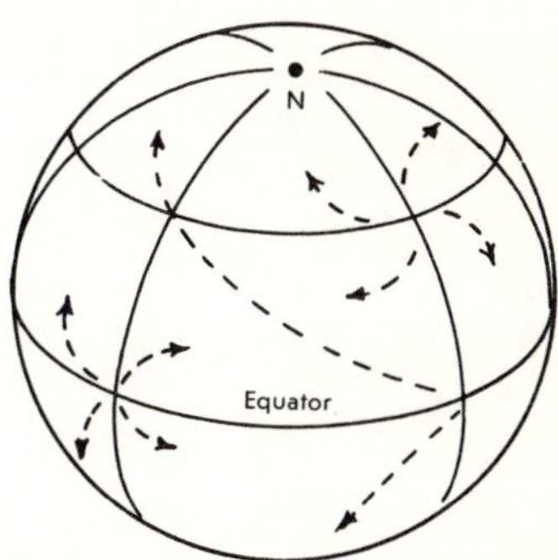

FIGURE 7.4. The Coriolis Effect

Southern Hemisphere, the small square will turn *clockwise* and the pendulum will be deflected *counterclockwise*.

In relation to the Coriolis Effect, the pendulum swinging in its plane compares to an object in motion across the earth's surface. Each is deflected from its earth grid direction as the earth turns beneath it. The object however, is not confined to one area as the pendulum is. It moves from one latitude to another. Consequently, as long as it is moving, that part of its deflection attributed to the Coriolis Effect varies as the sine of its changing latitude varies. Moreover, the more rapid the object's motion, the more rapid its latitude change. This influences the amount of deflection per unit of time. Thus, a slowly moving object is deflected less than a rapidly moving one at the same latitude, and an object at a high latitude will be deflected more than one at a low latitude if their velocities are the same. If it continues to move, the object will eventually follow a circular path, the so-called *circle of inertia*, if no other force acts upon it.

The Coriolis Effect is significant to both air and water movements on the earth's surface. Air moves from centers of high pressure to areas where pressures are somewhat lower. During this movement its otherwise straight path is curved, and it is to the curvature that we relate the whirling of air around air pressure cells. Furthermore, the directions of wind in the broad wind belts of the world are also affected. Similar deflection in the earth's great water bodies can be observed in the great drifts and ocean currents which characterize the Atlantic and Pacific Oceans. In the study of rivers, some students of the subject relate the winding paths of meandering streams to the Coriolis Effect.

Chapter **8**

Time and Its Measurement

Time is a basic conception which has to do with duration. It is measured by reference to events which occur in sequence and are repeated with regularity and accuracy. The earth's motions are such events, and we customarily refer our time to them. The day is the time required for the earth to make one rotation on its axis, and the year is the number of days required for it to make one revolution around the sun. How long a day is, of course, depends upon how fast the earth turns on its axis and just what one complete rotation is considered to be. Similarly, the length of a year will depend upon the earth's velocity in orbit as well as how a complete revolution is defined.

The day of 24 hours, or 86,400 seconds, is a basic unit in civil time, the system which is now in general use. It was long held to be the sole standard until the demands of modern science made it necessary to define a new standard which would take into account known changes in the earth's rotational velocity. Accordingly, in 1956, the *second* as a basic time unit was defined as a small fraction of a selected year. This gave the fundamental unit of *ephemeris time*[1] which is now used in scientific work. Other similar systems, also developed for scientific purposes, take into account minute and specific kinds of non-uniformity of motions during a particular year or years. Like ephemeris time, they are used where highly accurate determinations of time and/or position are to be made. All, however, are based upon long-established principles that are inherent in our civil time-keeping system and are discussed in the paragraphs that follow.

------

[1]The second of ephemeris time is defined as 1/31,556,925.9747 of the tropical year 1900.

Solar Time and Sidereal Time

The earth's motions are described relatively; that is, in their relations to other bodies in space. Time, too, is relative. Earth time may be measured by reference to earth movements as they relate to either the sun or the stars. In the case of the sun it is called *solar time.* In the case of a star it is *sidereal time.* Thus we speak of sidereal days and years and solar days. What might be thought of as a solar year is conventionally referred to as the *tropical year;* the time interval between two successive vernal equinoxes (March 21-22).

In making comparisons between solar and sidereal time it is necessary to take into account the apparent motions of the sun and the stars when they are observed from the earth. The earth moves and, from the viewpoint of the earthbound observer, the heavenly bodies appear to move. The apparent movements of the sun and stars depend upon two principal factors; their distances from the earth, and the nature of the "real" motions of the earth itself. Thus the earth-sun distance is 93,000,000 miles, almost inconceivably large. Yet, in terms of the dimensions of the universe, the distance is small enough so that earth movements impart to the sun two apparent motions in the sky. The sun appears to move from east to west during the day as the result of earth rotation, and from west to east as the year progresses as the result of earth revolution. In contrast, the nearest star is relatively so far away that it has but one apparent motion when observed by the unaided eye. Like the sun it appears to move westward as the result of rotation. Unlike the sun, however, its apparent position in the firmament is not affected by the earth's revolution. Even though two different positions of the earth in orbit may be as much as 186,000,000 miles apart (the length of the major axis of the orbit), this distance is relatively so small as to have no appreciable effect upon the observed position of a star at different times of the year unless very precise measurements are made. For most purposes a distant star may be regarded as being fixed in the sky, with the earth rotating under it. The star is the motionless body to which we refer when relatively precise statements of time are required.

When viewed from above the North Pole the earth moves in a counter-clockwise orbit around the sun. It makes one complete 360° revolution in one year of 365 1/4 days. This is to say that it moves through about 1° of its orbit per day.[2] This means that in relation to the earth, the sun appears to shift *eastward* at the rate of 1° per day as the result of revolution. At the same time the earth is rotating under the sun, also in

---

[2]Stated more precisely, this daily movement in orbit amounts to 360/365.25 degrees per day, if the earth moves at a constant velocity. For present purposes, however, 1° per day is considered to be an adequate approximation.

a counter-clockwise direction. It rotates in such a way and at such a rate as to give to the sun a second apparent daily motion of 360° *westward*. Taken together, the 1° eastward shift and the 360° westward shift give the sun an apparent total movement in the sky of 359° east to west during one complete period of the earth's rotation. This set of conditions is diagrammed in Figure 8.1. A first position of the earth in orbit (zero days) shows a reference meridian directly under the sun and a distant star. In the second position (the end of one sidereal day) the earth is shown after it has made one 360° rotation and has moved through about 1° of its orbit. In relation to the sun the total rotation of the earth during this period is clearly 359°. These relationships are the basis for distinguishing between the different kinds of days.

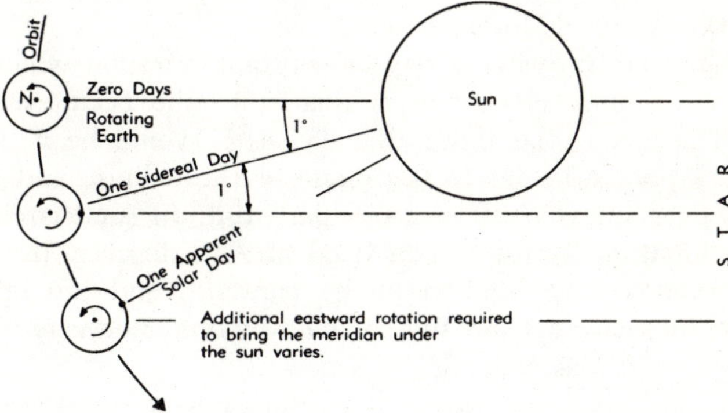

FIGURE 8.1. The Apparent Solar Day and the Sidereal Day

sidereal days and solar days

A day is the time interval between two successive passages of an earth meridian under a reference body. Stated differently, it is the time required for the earth to make one complete rotation in relation to either a star or the sun. As shown in Figure 8.1, when it is related to a star it is always equal to exactly one 360° rotational period, the time interval during which the earth moves about 1° along its orbital path. In contrast, when the day is related to the sun, it is slightly longer, by varying amounts, than one 360° rotational period. Figure 8.1 illustrates a situation within which the daily excess of rotation beyond 360° is more than one degree.

For most purposes, the rotational velocity of the earth can be considered to be a constant. Consequently, the day defined by reference to a star is also a constant. This is the *sidereal day*, which is divided into 24 hours of *sidereal time*. It differs from the *apparent solar day* which is not constant and which measures the time interval between two successive meridian passages under the sun. Variations in the apparent solar day relate to the inconstancy of the earth's orbital velocity and the daily change in the sun's declination. The apparent solar day is the day as shown on a sun dial.

mean solar days

Because the apparent solar day varies in length and the sidereal day is referenced to a star rather than to the sun and the occurrence of day and night, neither has general utility as a time unit. Consequently, for practical day-to-day time-keeping purposes, the *mean solar day* of 24 clock hours is used in *civil time*. This is the day which is meant when we say the year is 365 1/4 days long.

Mean solar time provides a day of constant duration which is measurable by a common clock and is related to the occurrence of day and night. The mean solar day is the day as it would be if the earth's axis remained perpendicular to the plane of the ecliptic and the earth moved at a constant velocity in a circular orbit. In such circumstances the sun's declination would remain fixed at zero degrees, the daily excess of rotation beyond 360° would be constant, and the relationship between the sidereal day and the solar day would not vary. This relationship is expressed as follows:

24 hours sidereal time = 23 hrs., 56 min., 4.09 sec., mean solar time
24 hours mean solar time = 24 hrs., 03 min., 56.555 sec., sidereal time

Variability of the Apparent Solar Day

With respect to variations in the apparent solar day, two principal factors are involved. These are the inconstancy of the earth's orbital velocity, which relates to the ellipticity of the orbit, and the continuing change in the sun's declination as the year progresses. The declination change results from the inclination of the earth's axis. Variations in rotational speed, although significant to very precise work, have little relevance to the more general problem of devising a functional time system for day-to-day use. Explanations at this time will then be confined to considerations of the orbital velocity and the declination change.

Considering only the orbital velocity of the earth, it will be recalled that the velocity is greatest when the earth is at perihelion, about January 6. The velocity is least when the earth is at aphelion, about July 4. This means that in January the earth moves slightly more than 1° per day in its orbit, and in July it moves slightly less than 1° per day. Therefore, the amount of rotation in excess of 360° which is necessary to bring the meridian back under the sun varies, depending on the time of year. The rotational excess is slightly larger in January than it is in July. For this reason the apparent solar day varies in length. It is longer in January than it is in July.

The effect of the varying orbital velocity is to lengthen and shorten the apparent day at the times of the December and June solstices respectively. This contrasts to the effects of the change in the sun's declination during the year. The declination changes because the axis is inclined. The rate of change, however, is not constant. It varies between the equinoxes and the solstices. The variation has the effect of lengthening the apparent day at both solstices and shortening it at the times of the equinoxes. Let us see how.

As the result of revolution, and disregarding the fact of rotation for the present, the sun's vertical ray and its subpoint follow a slanting path during their west-to-east movement from tropic to tropic during the period between two solstices. Continued through the next intersolstitial period the solar subpoint may be said to complete one journey around the earth as that body makes one complete passage through its orbit. The line it describes on the earth is a great circle which is inclined at an angle of 23°30′ to the Equator. This is shown in Figure 8.2.

The figure also shows segments of the subpoint's path at the times of the solstices and the equinoxes. The segments are of such lengths as to show the distances traveled by the subpoint in equal intervals of time. From the figure we can see that at the equinoxes the subpoint moves slantingly across the Equator; its due west-east movement is considerably smaller than its total movement and its west-east velocity along the Equator is decreased. This has the effect of *shortening* the time between two successive meridian passages under the sun. During the solstices, however, the total movement of the subpoint is nearly due west-east. Quite clearly the linear west-east movement of the subpoint at the solstices is more rapid than it is at the equinoxes. Consequently, in contrast to conditions at the equinoxes, at the times of the solstices the time interval between successive meridian passages is *lengthened*. This means that, as a result of the declination change alone, the length of the apparent solar day varies. The declination change tends to lengthen it at the solstices and to shorten it at the equinoxes.

There are, then, two principal factors which affect the length of the apparent solar day. Their combined effect is to sometimes make it longer or shorter than its corresponding mean day. More specifically, its

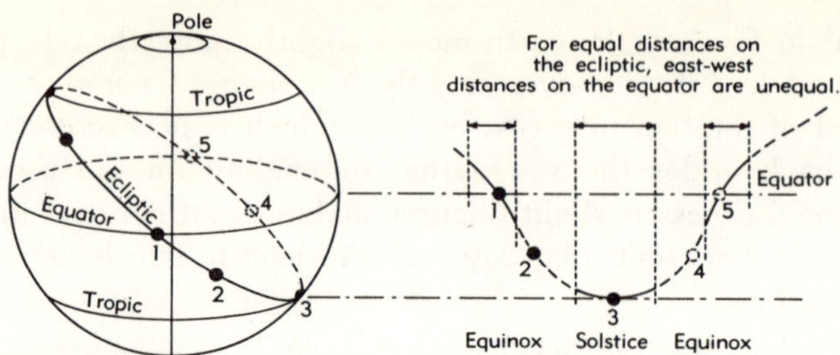

FIGURE 8.2.   The Path of the Solar Subpoint between the Solstices

length varies from 22 seconds shorter, to 28 seconds longer than the mean day. The differences between the two over a period of time however, are cumulative. This has the effect of creating considerable discrepancy between them at various times of the year. For example, during the months and weeks when the apparent day is the longer, the daily differences accumulate so that on February 12 the apparent day is about 14.5 minutes longer than the mean day. At the other extreme, on November 3, the apparent day is shorter by about 16.5 minutes. This condition has the interesting effect of increasing the total mean hours of sunshine during the months of the Southern Hemisphere summer. The increase is, of course, completely removed from the increase in hours of sunshine which relates to the season.

the equation of time

Quite clearly there can be no precise statement of equivalence between mean and apparent solar time. However, the two can be compared by referring to the equation of time which, briefly defined, is the difference between them on a particular day of the year. This difference can be observed, and the equation of time measured as follows. A highly accurate and stationary timepiece is set at noon, at precisely the time that the sun is overhead. At the following noon, 24 clock hours later, the position of the sun is compared with the position of the clock. If the sun is found to be east of the clock, it is said to be slow. That is, the apparent sun is slow in comparison with local mean time. If, on the other hand, the sun at clock noon lies to the west of the meridian, it is said to be fast.

the analemma

The equation of time and other related data are recorded on the analemma, the peculiar eight-shaped figure shown on many earth globes.

This is a graph which shows the equation of time and the sun's declination at any calendar day of a 365-day year. It has two axes. One, which extends north-south from tropic to tropic, may be thought of as a meridian. It is graduated into degrees and indicates the declination of the sun on various dates during the year. The second axis runs east-west along the Equator and is graduated in minutes which indicate the equation of time. Numbers on the right, east of the meridian, have negative values and indicate a slow sun (the clock is fast), while numbers on the left have positive values indicating a fast sun (with the clock slow). Points on the figure-eight curve are determined by plotting the number of minutes the sun is fast or slow against the sun's declination. The curve developed from them is graduated into the days and months of the calendar year.

In using the analemma, the sun's declination and the equation of time are read directly by observations of the intersections of the straight lines and the curve. For example, at the September equinox, the sun's declination is zero and the sun is fast by about 7.5 minutes. On August 19, the declination is 13° N, and the sun is slow by about 4 minutes. At four times during the year the curve intersects the north-south axis and apparent time is equivalent to mean time.

Because the analemma is graduated to show 365 days of the calendar year rather than the 365 1/4 days of the tropical year, it is accurate for only one year in four. When printed on a globe, however, it is a useful device for answering many questions about the sun's behavior at different times. For example, the curvature of the figure eight changes between the Equator and the tropics. The curves flatten noticeably in June and December as the result of the decreasing rate of daily declination change with the approach of the solstices. Moreover, the bulge of the curve is greater in the south than in the north. This demonstrates quite clearly the relatively high values of the equation of time during the Southern Hemisphere summer. It reflects the greater velocity of the earth in its orbit at the times when the earth is in or near perihelion, and contrasts to conditions associated with aphelion and the northern summer.

The Year

We have so far considered the year to be the time required for the earth to make one complete revolution around the sun; the interval between two successive vernal equinoxes. This is the year as it is defined in reference to the sun and the one from which our calendar is derived. It differs from the sidereal year which is the time interval between two

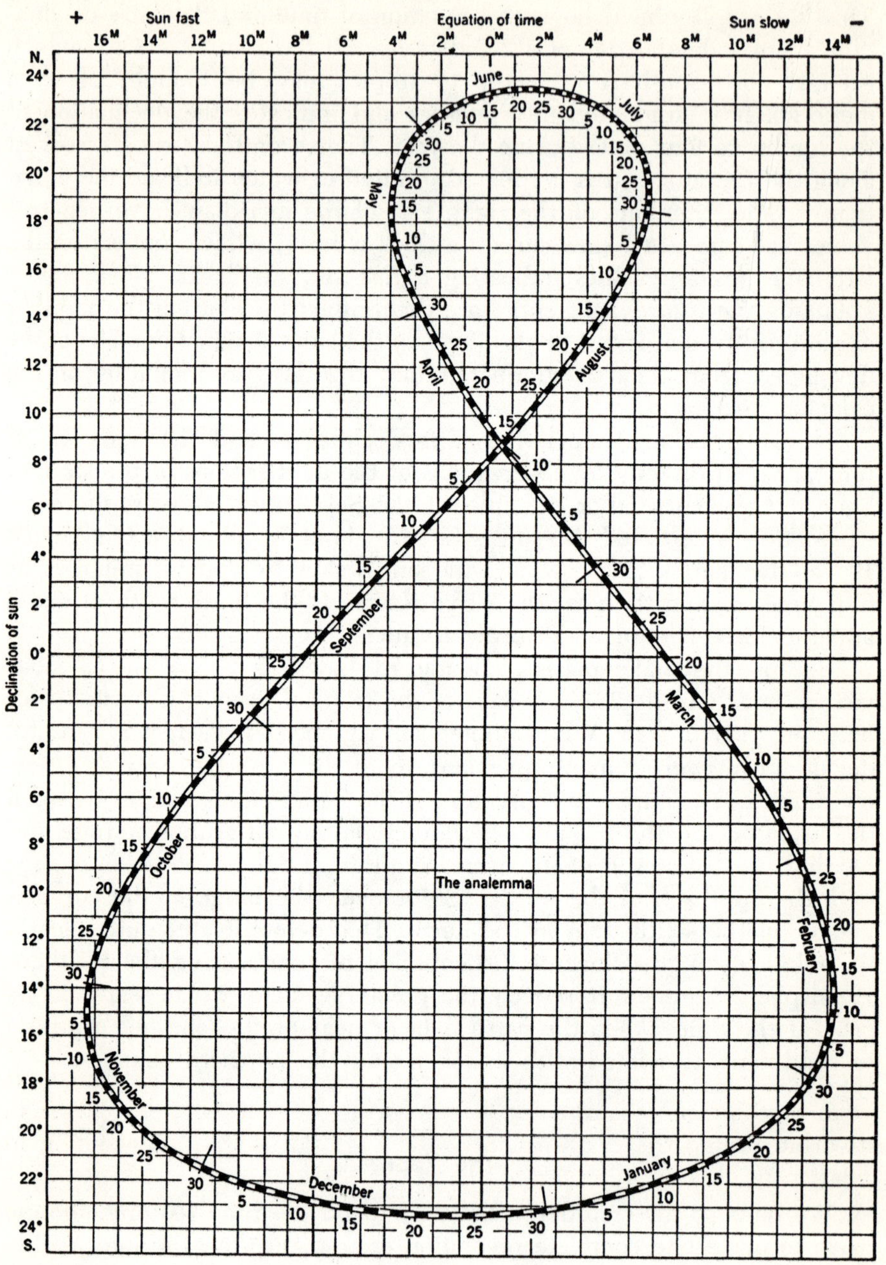

FIGURE 8.3.   The Analemma

successive appearances of the sun against the identical background of the stars and is about 20 minutes longer.[3]

As is the case with days, the year referenced to the stars has little relevance to the day-to-day problems of timekeeping. Accordingly, our discussion of the year is confined to considerations of the tropical year, a time interval which is more closely related to the calendar and the seasons and is therefore more generally functional.

The length of the tropical year is 365 1/4 days or, more precisely, 365 days, 5 hours, 48 minutes and 46.08 seconds, of mean solar time, a difference of about 11.25 minutes a year, or one day in 128 years. On the other hand, the calendar contains 365 days and, in leap year, three hundred and sixty-six. This arrangement corrects the otherwise lost one-quarter day when the less precise definition of the year is used. It does not, however, make a complete correction such as would be required when the more precise definition of the year is taken into account. The more complete correction is required to prevent excessively increasing disparities between calendar dates and seasons. That is, the calendar and solar years must be made to correspond as closely as possible for as long a time as possible. Correction is made by considering century years as leap years only when they are divisible by four hundred. Thus the year 1600 was a leap year and had 366 days. In contrast, the years 1700, 1800, and 1900 were also leap years but, since they were not divisible by 400, the extra day was not added. The year 2000 will be a leap year of 366 days. Such adjustment reduces the average deviation of the calendar year from the solar year to about 26 seconds, or less than a day in some 3000 years. Thus the critical dates of the solstices and the equinoxes are practically constant; seasons and dates during different years correspond with reasonable accuracy, and the calendar year is nearly identical with the solar year over a period of several centuries.

The Month

As a unit of time the month may be considered in one of two ways: as a convenient grouping of days indicating about 1/12 of a calendar year, or the time required for the moon to complete one revolution in its orbit around the earth. In the latter case, we define a revolution in relation to either the sun or a star.

Observed from a point above the earth's North Pole, the moon moves in a counter-clockwise direction in its orbit. It makes one 360° revolution

---

[3]It should be recalled here that precession causes the equinoxes to shift westward against the apparent eastward 1° daily shift of the sun which relates to revolution. This has the effect of shortening the expected tropical year by about 20 minutes.

in 27.32 days, a time interval known as the *sidereal month,* or the period of the moon's revolution in terms of the stars. Stated differently, if observed from the earth, the moon will be seen to advance eastward 360° every 27.32 days, or about 13.2° each day in relation to some distant star reference. At this juncture it will be recalled that the advance is in the same direction as the apparent mean daily shift of the sun which relates to the earth's revolution. From this it follows that the moon's eastward movement in reference to the sun is approximately 12.2° per day: 13.2° minus the sun's daily change of 1°. This means that, in relation to the sun, the moon makes one complete revolution of the earth in 29.5 days, or one *synodic month.* The figure 29.5 is arrived at by dividing 360° by the daily movement of 12.2°.

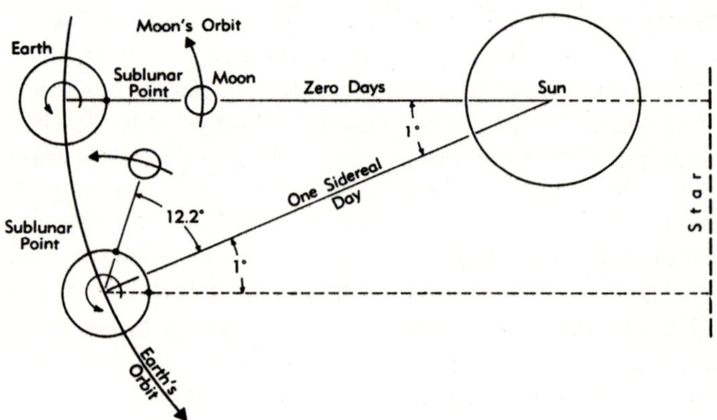

FIGURE 8.4.   Earth-Sun-Moon Relationships

Clock Time and Longitude

As previously noted, clock time derives from considerations of the movement of the rotating earth through an hypothetical circular orbit, the plane of which passes through the Equator. In such an orbit the path of the sun's subpoint would at all times correspond with the Equator. Since in this hypothetical situation the rotational and orbital velocities of the earth are unvarying, the sun appears to move through the sky at an unchanging rate of speed. The noon-to-noon time periods at any meridian and at any time of the year are equal in length, as are the periods of time required for the sun to pass over any longitude intervals which are equivalent to one another. Moreover, the total apparent movement of the sun around the earth in one day is exactly 360° of longitude. Since the time required for the movement is arbitrarily fixed at 24 mean

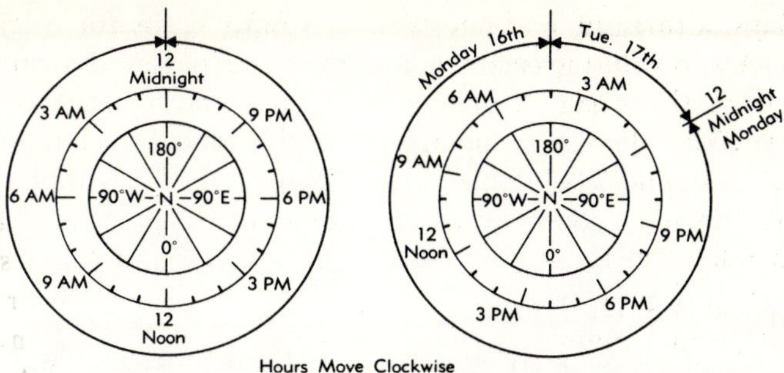

FIGURE 8.5. Longitude-Time Relationships

solar hours, one hour of mean solar time is the time required for the apparent movement through 15° of longitude.

If viewed from above the North Pole, the earth rotates in a counter-clockwise direction under the sun. The sun, then, has an apparent movement clockwise around the earth, as does the noon hour. All other hours move with it, with the midnight hour being at all times located 180° away, on the opposite side of the globe. Six A.M. (from *Ante Meridian,* meaning before the meridian passage) is 90° "ahead" of it, while 6 P.M. (from *Post Meridian,* meaning after the meridian passage) is 90° "behind" it. Hour changes may then be thought of as deriving from the movement of a clock dial clockwise around the earth when that body is observed from above the North Pole. As the various hour markers on the dial move across a meridian the hours follow one another in their proper sequence.

With respect to changes in the days of the week, one day begins and another ends at the midnight hour, as shown on the clock dial. Quite obviously, however, the day must also change at some other place as well, to avoid the confusion that would result from an indeterminate merging of one day with its successor. This place has been arbitrarily selected as the 180th meridian, the International Date Line. Here the days change as follows. If one travels between the Western and Eastern Hemisphere (as in going from North America to China) the day would be advanced from, let us say, the sixteenth of the month to the seventeenth upon crossing the Date Line. In the opposite sense, when the crossing is made while traveling from China to North America, the day changes from the seventeenth to the sixteenth. In the meanwhile, hourly changes take place systematically as described above.

From the foregoing it is evident that for most of the time there are two calendar days in existence. More than two is an impossibility. However, when it is noon at Greenwich and the midnight hour is at the

International Date Line, one day, and one only, exists for a fraction of a second. Immediately after the midnight hour passes the line, a new day is begun and, as that hour sweeps around the globe, the new day covers an increasingly large increment of the global surface while the area which is experiencing the older previous day is shrinking correspondingly.

time zones and local time

We have so far discussed clock time as local time; that is, time as it varies from meridian to meridian. Since every place on the earth has its own meridian it also has its own time, and any two places on the earth which have the same longitude must have the same time. However, time reckoned in this manner is difficult to work with simply because if it were in use all over, any east-west or west-east movement would necessarily involve a resetting of the clock. Clock time is therefore generalized and said to be equal within broad longitudinal belts called Time Zones. Mathematically there are 24 such zones, with each centered on a Standard Meridian and identified by reference to it. Each zone keeps the time of its standard meridian and clocks need to be reset only when one is traveling from one zone to another.

The standard meridians, beginning at Greenwich, are each 15° of longitude apart, with the time zones extending 7°30′ to the east and the west. The Pacific Standard Time Zone which is centered on 120° W longitude should, in theory at least, extend from 127°30′ W to 112°30′ W. At Greenwich, the zone extends from 7°30′ E to 7°30′ W, and at the International Date Line from 172°30′ E to 172°30′ W. In practice, the zone boundaries are often modified to suit local requirements and in consequence the world map of time belts shows considerable variation from the ideal.

World Standard Time is time as it is established by the time zones. It was adopted at the International Meridian Conference in Washington, D.C., in 1884, and has served to regularize the time relationships between most parts of the world since that time. Although some countries were slow in adopting it most of the world's nations have by now accepted the time zone principle as outlined above. Notable exceptions are some of the Asian countries. Where the principle has been adopted, and deviations from the 15° belts occur, they can usually be explained by reference to considerations given to such factors as political boundaries, difficulties in scheduling long distance transport, local convenience, and the like. A notable case is that of China. Although that country extends through some 50° of longitude, its time is standardized through-

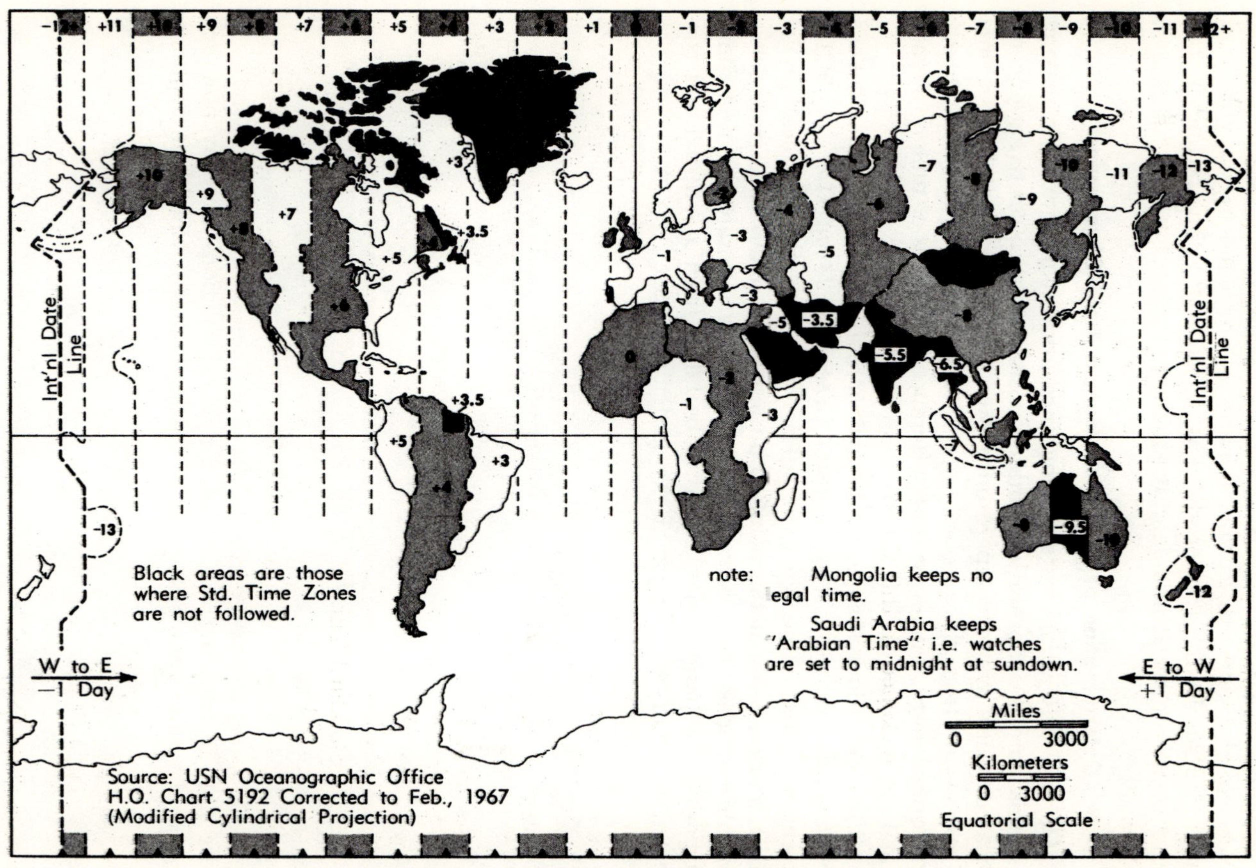

FIGURE 8.6. World Time Zones Generalized

out to conform with the meridian of 120° E. In the case of the U.S.S.R. and countries in Central Europe the arrangement of time zones is the reverse of the standard. Further departures are noted in Mongolia and Saudi Arabia. In the former there is no standard time, while in the latter clocks are set at midnight at sundown.

Determination of Longitude from a Time Observation

Quite clearly a time system as outlined above can be used in making a simple determination of longitudinal position. It is a relatively uncomplicated procedure to compare one's own time with time at the Prime Meridian and from this to obtain one's longitude. Since 15° of longitude are equivalent to a one-hour time difference, 4 minutes of time difference equals 1° of longitude difference and so on.

Prime Meridian time is read from a highly accurate timepiece (a *chronometer*) which is set to register *Greenwich Mean Time,* (G.M.T.) or obtained from radio signals broadcast from any one of several official observatories. Time at the point to be located is determined from celestial observations and reference tables. For less accurate determinations of position, apparent solar time may be taken from a sun dial and corrected to mean time by use of the analemma. (See App. A.)

Appendix A

Determination of Geographic Position

The approximate geographic coordinates of an observer can be determined by the following method. Figures A.1 and A.2 are given in illustration.

Given: These data are obtained from the calendar and the clock, and by observations and measurements made by the observer.

1. The date is May 16.
2. Apparent Solar Time (AST) is 12 noon. The sun is directly above the observer's meridian.
3. The sun's altitude is 30° N. The angle is measured between the observer's horizon and his line of sight to the sun. The observer looks north to see the sun, (Figure A.1). This measurement must be made at 12 noon, AST.
4. Greenwich Mean Time (GMT) or local Mean Time at the Prime Meridian is 4 AM. This time is read from a clock or is received by radio.

The Solution: The determination of the position of the observer is carried out in two steps. In solving for latitude, Figure A.1 is given in illustration. Figure A.2 illustrates the longitude solution. (See page 112.)

Solving for The Latitude: The observer is first located in relation to the vertical sun ray by determining the latitude difference between them. The sun's declination is then applied to the difference to determine the observer's position in relation to the equator. Figure A.1 illustrates Case 3 described on page 89.

1. The latitude difference and the zenith angle are equal. In our case the difference is 60°, or 90° minus the sun's altitude which is given (measured) as 30°. In explanation, in Figure A.1 the line drawn

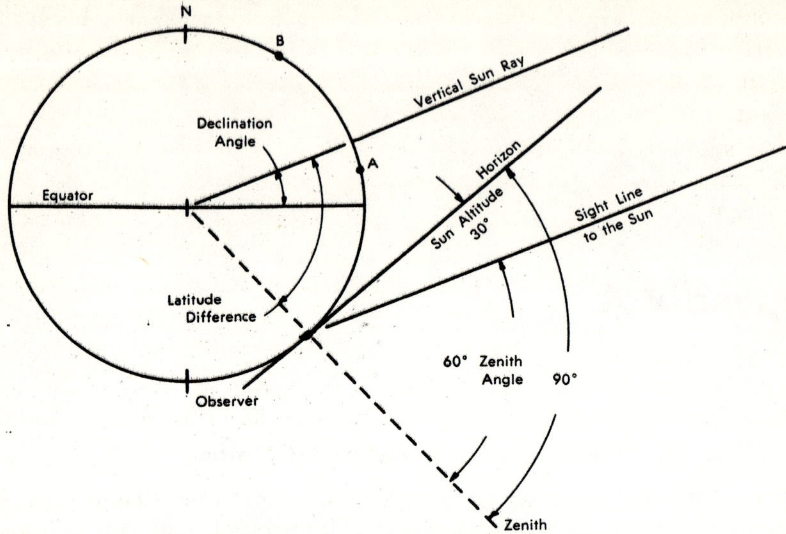

FIGURE A.1. Latitude Determination

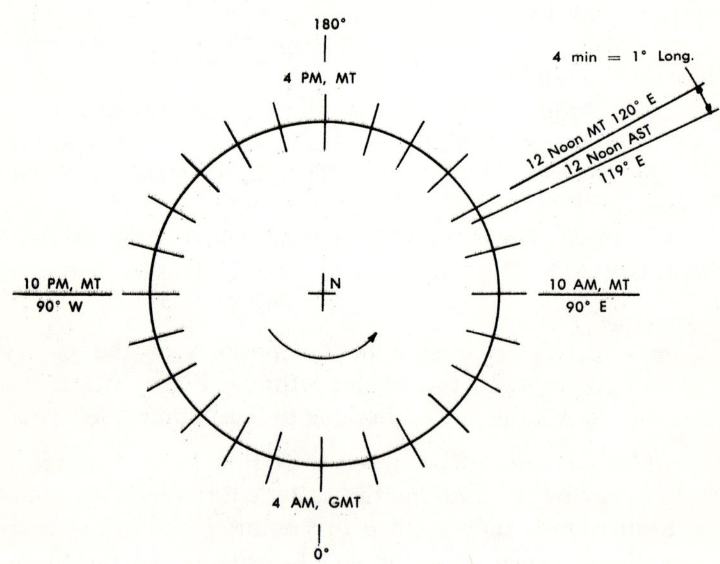

FIGURE A.2. Longitude Determination

from the zenith to the earth's center is a transversal which cuts two parallel lines, the vertical sun ray and the sight line to the sun which is an oblique ray. If two parallel lines are cut by a transversal, the exterior-interior angles are equal.
2. The sun's declination on May 16 is 19° N. (See the Analemma, Figure 8.3).
3. Subtracting the declination angle, 19° N, from the latitude difference, the observer's latitude is 41° S.

Note: If the observer were located between the equator and the vertical ray, at some point A, the latitude difference would be subtracted from the declination angle, (Case 2, page 89). If the observer were located north of the vertical ray, at some point B, the latitude difference would be added to the declination angle, (Case 1, page 89).

Solving for the Longitude: The observer calculates the Mean Time (MT) difference between his own position and GMT. Time difference is converted to longitude difference by means of the conversion—one hour of MT equals 15° of longitude. Twelve o'clock noon, AST, is determined by a sun observation and converted to MT by use of the equation of time.

1. On May 16, the sun is fast by 4 minutes. This means that the sun is "ahead" of the MT noon hour by 1° of longitude.
2. Since GMT is given as 4 AM, the MT noon is at longitude 120° E.
3. The observer is located at 119° E.

Note: If the sun were slow by 4 minutes, the observer's position would be at 121° E, "behind" the MT noon hour.

Bibliography

Burkhard, R. K., Capt. U.S.A.F. *Geodesy for The Layman*, U.S.A.F. Aeronautical Chart and Information Center. St. Louis, Missouri, 1962.

Deetz, C. H., and Adams, O. S. *Elements of Map Projections, with Applications to Map and Chart Production*, U.S. Dept. of Commerce, Coast and Geodetic Survey, Spec. Pub. #68, 5th ed. U.S. Gov't. Printing Office, Washington, D.C., 1945.

Dept. of the Army, Field Manual FM 21-26, *Map Reading*. U.S. Gov't. Printing Office, Washington, D.C. 1969.

Garland, G. D. *The Earth's Shape and Gravity*, New York: Pergamon Press, 1965.

Greenhood, D. *Mapping*, Chicago: University of Chicago Press, 1964.

Harrison, L. C. *Sun, Earth, Time and Man*, Chicago: Rand McNally & Co., 1960.

Kellaway, T. G. *A Background of Physical Geography*, London: The Macmillan Company, 1960.

Mehlin, T. G. *Astronomy*, New York: John Wiley & Sons, 1961.

Mitchell, H. C. and Simmons, L. G. *The State Coordinate Systems, A Manual for Surveyors*, U.S. Dept. of Commerce, Coast and Geodetic Survey, Spec. Pub. #235, U.S. Gov't. Printing Office, Washington, D.C., 1945.

Raisz, E. *Principles of Cartography*, New York, San Francisco, Toronto, and London: McGraw-Hill Book Co., Inc., 1962.

Ryabov, Y. *An Elementary Survey of Celestial Mechanics*, translated from the Russian by G. Yankovsky. New York: Dover Pub., 1961.

Strahler, A. N. *Physical Geography*, 2nd edition, New York: John Wiley & Sons, Inc. 1960.

——— *The Earth Sciences*, New York: Harper & Row, Publishers, 1963.

Stumpff, K. *Planet Earth*, Ann Arbor: Univ. of Mich. Press, 1959.

Dept. of the Air Force, *A.F. Manual 51-40, vol. I, Air Navigation*. U.S. Gov't. Printing Office, Washington, D.C., 1962.

Index